Fighting the Opioid Epidemic

The Role of Providers and the Clinical Laboratory in Understanding Who is Vulnerable

Fighting the Opioid Epidemic

The Role of Providers and the Clinical Laboratory in Understanding Who is Vulnerable

AMITAVA DASGUPTA

Professor, Pathology and Laboratory Medicine,
McGovern Medical School,
The University of Texas,
Houston, TX,
United States

ELSEVIER

Elsevier
Radarweg 29, PO Box 211, 1000 AE Amsterdam, Netherlands
The Boulevard, Langford Lane, Kidlington, Oxford OX5 1GB, United Kingdom
50 Hampshire Street, 5th Floor, Cambridge, MA 02139, United States

Notices
Knowledge and best practice in this field are constantly changing. As new research and experience broaden our
understanding, changes in research methods, professional practices, or medical treatment may become
necessary.

Practitioners and researchers must always rely on their own experience and knowledge in evaluating and using
any information, methods, compounds, or experiments described herein. In using such information or methods
they should be mindful of their own safety and the safety of others, including parties for whom they have a
professional responsibility.

To the fullest extent of the law, neither the Publisher nor the authors, contributors, or editors, assume any
liability for any injury and/or damage to persons or property as a matter of products liability, negligence or
otherwise, or from any use or operation of any methods, products, instructions, or ideas contained in the
material herein.

Library of Congress Cataloging-in-Publication Data
A catalog record for this book is available from the Library of Congress

British Library Cataloguing-in-Publication Data
A catalogue record for this book is available from the British Library

ISBN: 978-0-12-820075-9

For information on all Elsevier publications visit our website at
https://www.elsevier.com/books-and-journals

Publisher: Stacy Masucci
Acquisitions Editor: Stacy Masucci
Editorial Project Manager: Megan Ashdown
Production Project Manager: Sreejith Viswanathan
Cover Designer: Greg Harris

Typeset by TNQ Technologies

Working together
to grow libraries in
developing countries

www.elsevier.com • www.bookaid.org

For Thor and Minnie

Preface

Various opioid drugs are used in pain management, but unfortunately nonprescription abuse of opioid is becoming a serious public health issue. On October 26, 2017, Acting Health and Human Services (HHS) secretary Eric D. Hargan declares opioid crisis as a public health emergency. Therefore, identifying population who are vulnerable to alcohol and drug addiction is important to fight opioid epidemic. There is no single genetic polymorphism, which is associated with alcohol and drug abuse including opioid abuse, but it has been recognized that polymorphisms of multiple genes contribute collectively in increasing risk of alcohol and drug addiction. In general, genetics contribute approximately 50%, whereas environmental factors contribute 50% in increasing risk of alcohol and drug addiction in an individual. The toxicology laboratory also can play a vital role in fighting opioid epidemic by implementing a robust system for drugs of abuse testing as well as drug testing in pain management patients. Not detecting intended drug in urine indicates noncompliance or selling such drug in clandestine market. In addition, it is also important to identify patients who abuse other illicit drugs along with prescription opioid. Moreover, opioid abuse may cause medical emergency where identification of alcohol and any other drugs along with opioid is essential. However, common pitfalls of pain management testing must be avoided by communicating with ordering physician. A common mistake is negative drug testing result in a patient taking keto-opioids such as oxycodone and oxymorphone because the clinician ordered opiate drug screening in urine where immunoassay antibody designed to detect morphine lacks sufficient cross-reactivity with keto-opioids for detecting therapeutic levels of such drugs in urine after proper use. Similarly, other commonly used opioids such as oxymorphone, tramadol, buprenorphine, methadone, and fentanyl cannot be detected by opiate immunoassay. Therefore, to monitor pain management patients, it is important to implement proper testing and policies in the toxicology laboratory including capability of confirming drugs using gas chromatography/mass spectrometry or liquid chromatography/tandem mass spectrometry. This book covers all important aspects essential to fight opioid epidemic.

This book has seven chapters. In Chapter 1, brief history of opium abuse and current abuse of heroin are discussed. In Chapter 2, an overview of prescription opioids is presented, while in Chapter 3, opioid abuse and opioid epidemic are discussed in detail. In Chapters 4 and 5, genetic factors and environmental factors that contribute to susceptibility of opioids as well as substance abuse are addressed, respectively. Some people use poppy seed defense to justify heroin abuse. Factors that can differentiate between poppy seed use and heroin abuse are discussed in detail in Chapter 6, and finally in Chapter 7, drug testing in pain management is addressed along with its pitfalls.

This book is written with an objective to have a broad range appeal to physicians practicing pain management, psychiatrists dealing with drug and alcohol rehabilitation, pathologists, toxicologists, and clinical scientists involved in drug testing in pain management, resident physicians, senior medical students, nurse practitioners, and social workers. I want to acknowledge support of my wife Alice while I spend long evening hours and weekends to write this book, which took me 1 year. If readers enjoy this book, my efforts will be duly rewarded.

Respectfully submitted
Amitava Dasgupta, PhD
Houston, TX

Contents

CHAPTER 1

Crude Opium, Morphine, Codeine, and Heroin: Past and Present

INTRODUCTION

Poppy plants (*Papaver somniferum*, *Papaver paeoniflorum*, and *Papaver giganteum*) are herbaceous annual plants that can grow almost anywhere. Of all the different species, *P. somniferum* is the most popular plant due to its beautiful flowers as well as its seeds, which are used for making muffins, baking breads, and other purpose. This plant is one of the oldest medicinal plants known to mankind. The plant's seedpod contains a latex solution that exudes if scored with a sharp instrument. The latex is raw opium that contains both alkaloids and nonalkaloids [1]. Crude opium was used for various purposes since ancient time. Most likely, opium was the first narcotic substance discovered at the dawn of mankind [2].

OPIUM USE: FROM ANCIENT TIME TO MODERN TIME

Opium use was known to ancient cultures as early as 3500 BC by Sumerians who called it "Hul Gil," which means "joy plant." The Sumerians passed the knowledge of poppy cultivation to the Assyrians, the Babylonians, and ultimately, the Egyptians. The description found on the Ebers Papyrus (1500 BC) from Egypt indicated use of poppy seeds as a remedy for pain relief as well to prevent excessive crying in children. Interestingly, artifactual remains related to opium usage were also found in the Egyptian tomb from 15th century BC. A goddess from about 1500 BC showed her hair adorned probably with poppy capsules and her closed eyes were indication of sedation, thus speculating opium use in ancient time. In addition, juglets probably imitating the poppy capsules were found in that period in both Cyprus and Egypt. By 1300 B.C., the Egyptians were cultivating opium and probably traded opium all over the Middle East and into Europe [3]. Around 850 BCE, the great Greek author Homer considered opium as an intoxicating substance with pain-relieving and sleep-inducing properties [4].

Hippocrates (460–375 BC), the father of medicine, recognized analgesic effects of opium and probably prescribed poppy juice as a purgative, a narcotic, and possibly a cure for leucorrhea. Around 330 B.C., Alexander the Great introduced opium to the people of Persia and India, where the poppies later came to be grown in vast quantities. The first recorded reference of utilizing juice from poppy as a therapeutic agent was made by Theophrastus, the Greek scholar in the third century BC. In the second century, AD Roman physician Galen introduced use of opium for medical use. However, he was also aware of abuse potential of opium [4]. The antidiarrheal property of opium was mentioned by Persian physician Avicenna (980–1037 AD). He also reported that use of opium common in Khorasan (northeast of Iran) and Bokhara (southwest of Uzbekistan). Arab traders introduced opium to India and China during later part of Tang dynasty (618–907 AD). Probably between 10th and 13th century, opium was introduced in the Asia Minor and then to Europe [4].

Ingestion of opium as an oral preparation has undergone many modifications over the Centuries. Thomas Dover (1660–742), an English physician, developed a powdered form of edible opium that was also known to contain saltpeter (mostly potassium nitrate), tartar, licorice, and ipecacuanha. Laudanum, an alcoholic tincture of opium, was created by the 17th century English physician, Thomas Sydenham, and was commonly paired with whiskey or rum. As a means for preparing patients for surgery, Laudanum became widely used in Europe and North America into early 20th century. Moreover, one of the first books dedicated to the uses and effects of opium, *Mysteries of Opium Reveal'd*, written by the Welsh physician, John Jones, was published in 1701. However, he also listed many side effects of using opium [5].

Opium was used for baby sedation from the ancient time. As a free rider of gum lancing, opiates joined the treatment of difficult teething in the 17th century. During Industrial Revolution, opium was used by working class. With industrialization, private use was rampant among the working class. In German-speaking countries, poppy extracts were administered in soups and pacifiers. In English-speaking countries, proprietary drugs containing opium were marketed under names such as soothers, nostrums, anodynes, cordials, preservatives, and specifics and sold at the doorstep or in grocery stores. Opium's toxicity for infants was common knowledge; thousands of cases of lethal intoxication had been reported from antiquity, but physicians continued to prescribe it for babies. Unregulated trade led to greatly increased private use of opiates during the 19th century. Intoxication became a significant factor in infant mortality. As late as 1912, the International Hague Convention forced governments to implement legislation that effectively curtailed access to opium and to discontinue the dangerous habit of sedating infants [6].

Before the development of general anesthesia, surgery was performed only in extreme necessity. It is probable that an analgesic such as opium would have been given during or following surgery. The first description of opium usage for postoperative analgesia is dated to 1784, when the Glasgow-born London surgeon, James Moore, described use of opium. The Scottish surgeon, Benjamin Bell, also noted that opium was useful in controlling postsurgical pain [5].

According to historical evidence, the abuse of opium has been reported all over the globe—specifically throughout Eastern nations since the 16th century. Before that, opium was mostly used as medicine. Reference has been made in traditional Persian medical literature to the method of cultivation, properties, side effects, and toxicity. In 16th century Iran, opium use began during the reign of the Safavids. Then famous scholar Imad al-Din Mahmud ibn Mas'ud Shirazi composed a book concerning addiction, *Afyunieh*, a comprehensive book on the topic of opium and all issues of opium abuse. Furthermore, he recommended methods for reducing opium dose as well as substitution with other medications that had a narrower range of side effects, in order to eradicate dependency upon opium and opium-derived materials. This is most likely the first book that comprehensively addressed opium and discussed drug rehabilitation methodology, in traditional Persian medical literature [7].

Along with opium use, opium addictions also become a significant health hazard. Opium addiction was widespread in China, and following a 1799 ban on opium in China, smuggling of opium became a common industry that lead to the opium war between the British and China in 1839. During the 19th century,

opium was grown in the United States as well as imported. Besides indiscriminate medical use, opiates were available in the United States in myriad tonics and patented medicines. In addition, smoking in opium dens was unhindered, resulting in an epidemic of opiate addiction by the late 1800s. The development of hypodermic syringe in 1853 by Alexander wood unfortunately pushed opiate abuse including heroin abuse to a new level. The decade before Civil War, United States was armed with most powerful painkiller known to mankind at that time which was morphine. Morphine could be easily injected into the body using a hypodermic syringe. At that time, opiate-addicted American population was mainly consisted of Caucasian women who legally purchased opium-laced cough syrup at local pharmacies or who obtained "laudanum," a tincture of opium by mail. Such addiction was virtually invisible because most of these women stayed at home. The generous use of morphine in treating wounded soldiers during the Civil War in the United States also produced many addicts. Surveys between 1878 and 1885 indicated that 56%—71% of opiate addicts in the United States were middle- to upper-class white women who purchased the drug legally. Heroin was first synthesized in 1874. Heroin was considered as a highly effective medicine for treating cough, chest pains, and discomfort of tuberculosis [8].

Heroin and opioid abuse is now a global public health issue. According to the word drug report of 2014, the estimated total amount of opium produced worldwide was 5500 tons with the overwhelming majority of production continued to take place in Afghanistan (80% of total production) and Myanmar (10% of total production) [9].

OPIUM AND POPPY PLANTS: LEGAL ISSUES

In 1875, the City of San Francisco adopted an ordinance in order to prohibit smoking of opium. In 1905, the United States Congress banned opium, and in 1906, the Pure Food and Drug Act required accurate labeling of all patented medicines. The Harrison Narcotics Act of 1914 taxed and regulated the sale of narcotics and prohibited giving maintenance doses to addicts who made no effort to recover, leading to the arrest of some physicians and closing of maintenance treatment clinics [10]. Since then, numerous laws attempting to regulate importation, availability, use, and treatment have been passed. Opium and its constituent chemicals are listed as Schedule II drugs, while heroin is a Schedule I drug. Opium poppy for legal pharmaceutical purposes is grown in various countries under government license, but very little is produced in the United States. The countries that can legally import crude opium to the United States include India, Turkey, Spain, France, Poland, Hungary, and Australia. India is the largest importer of opium to legal pharmaceutical markets of the world. Unfortunately, large quantities of opium are still grown worldwide mostly for use in the illegal manufacture of heroin. The opium gum may be crudely refined and smoked (e.g., "brown sugar") or converted to morphine and heroin.

Growing opium poppies (*P. somniferum*) are currently illegal in the United States. The sale of poppy seed from *P. somniferum* is banned in Singapore and Saudi Arabia.

ALKALOIDS FOUND IN OPIUM

Opium is a complex mixture of nonalkaloids and alkaloids. The nonalkaloid parts include water (5%–20%), sugar (20%) and several organic acid, most common being meconic acid (3%–5%). More than 40 different alkaloids have been isolated from opium. However, major alkaloids are morphine (4%–21%), codeine (0.8%–2.5%), thebaine (0.5%–2.0%), papaverine (0.5%–2.5%), noscapine (narcotine: 4%–8%), and narceine (0.1%–2%) [4–9]. Minor alkaloids present in opium include oripavine, boldine, reticuline, codamine, laudanidine, laudanosine, norlaudanosine, and cryptopine.

Identification of specific opium alkaloids in the seized opium samples is useful not only for judicial purpose but also to locate geographic location of cultivation of opium and trafficking of opium. The morphine content in opium is a key variable in calculations to estimate global illicit supply of heroin. Remberg et al. analyzed 78 opium samples collected from different regions in Afghanistan during the years 2000–05 to establish possible correlations between alkaloid contents and selected external factors, such as region of origin, year of harvest, use of fertilizer, extent of irrigation, or intrabatch variation. The authors analyzed morphine, codeine, papaverine, and thebaine in opium samples using high-performance liquid chromatography. The average morphine content was 14.4% in 78 specimens. Mean and range of alkaloids content in opium specimens analyzed by the authors are listed in Table 1.1. The authors observed higher alkaloid content when poppy plants were cultivated during warmer seasons. Using 14.4%, the average morphine content of all 78 samples analyzed, and assuming an overall laboratory extraction efficiency of

TABLE 1.1 Mean and Range of Alkaloids Content of 78 Opium Specimens Analyzed by Remberg et al.		
Opium Alkaloid	**Mean Alkaloids Content**	**Range**
Morphine	14.4%	5.1%–24.8%
Codeine	5.9%	2.6%–9.7%
Thebaine	4.0%	0.7%–9.3%
Papaverine	2.8%	0.0%–9.3%

Source of data: Reference Remberg B, Sterrantino AF, Artner R, Janitsch C, Krenn L. Science in drug control: the alkaloid content of afghan opium. Chem Biodivers. 2008;5:1770–1779.

70%, 10 kg of dry opium could be converted into about 1 kg of pure heroin base (i.e., the traditionally used 10:1 rule of thumb for opium-to-heroin conversions) [11].

Narayanaswami et al. analyzed 28 opium specimens collected from different geographic locations and analyzed them using gas chromatography after converting them to silyl ethers. The authors concluded that Japanese sample had the highest morphine content and the lowest narcotine content when compared with the samples from nine different geographical regions. Ecuador samples were low in codeine and thebaine while high in narcotine. The Mongolian sample had the lowest morphine content. However, highest morphine content was observed in one specimen collected from Indian state of Madhya Pradesh [12]. Alkaloid contents of opium collected from different countries are listed in Table 1.2.

MORPHINE

Morphine (from Morpheus, the Greek god of dreams) is the first alkaloid that was isolated from opium in 1817. Chemical structure of morphine is given in Fig. 1.1. Morphine is probably the most potent alkaloid present in opium because it is responsible for analgesic, euphoric, sedative, and also addictive behavior. However, morphine is also responsible for many adverse effects of opium. During 1830s, morphine became very common analgesic, and after American Civil War, many wounded veterans were addicted to morphine, to the extent that the name "soldier's disease" was coined to describe dependence of soldiers on morphine [4]. At present, crude opium is only used as a raw material for synthesis of various opioids. Morphine is widely used today as a narcotic analgesic, but it is also abused. Medical use of morphine is discussed in Chapter 2.

TABLE 1.2
Alkaloid Content of Opium Collected From Different Countries and Measured by Gas Chromatography.

Country	Morphine	Codeine	Thebaine	Papaverine	Narcotine
Inner Mongolia	6.97%	4.88%	1.29%	0.93%	6.62%
Manchuria	9.25%	4.69%	2.75%	1.22%	5.43%
China	10.13%	6.47%	1.64%	1.27%	8.79%
Japan	17.85%	2.32%	1.05%	2.17%	3.47%
Korea	11.27%	3.92%	1.74%	0.70%	4.65%
Yugoslavia	11.28%	5.58%	0.48%	2.73%	6.30%
Greece	16.9%	1.71%	0.5%	3.31%	8.8%
Turkey	14.02%	2.05%	0.84%	1.65%	5.82%
Equator	13.81%	1.71%	0.25%	2.62%	10.17%
India (Madhya Pradesh, neemuch II)	18.94%	2.31%	0.64%	1.11%	4.99%

Source of data: reference Narayanaswami K, Golani HC, Dua RD. Assay of major and minor constituents of opium samples and studies of their origin. Forensic Sci Int. 1979;14:181−190.

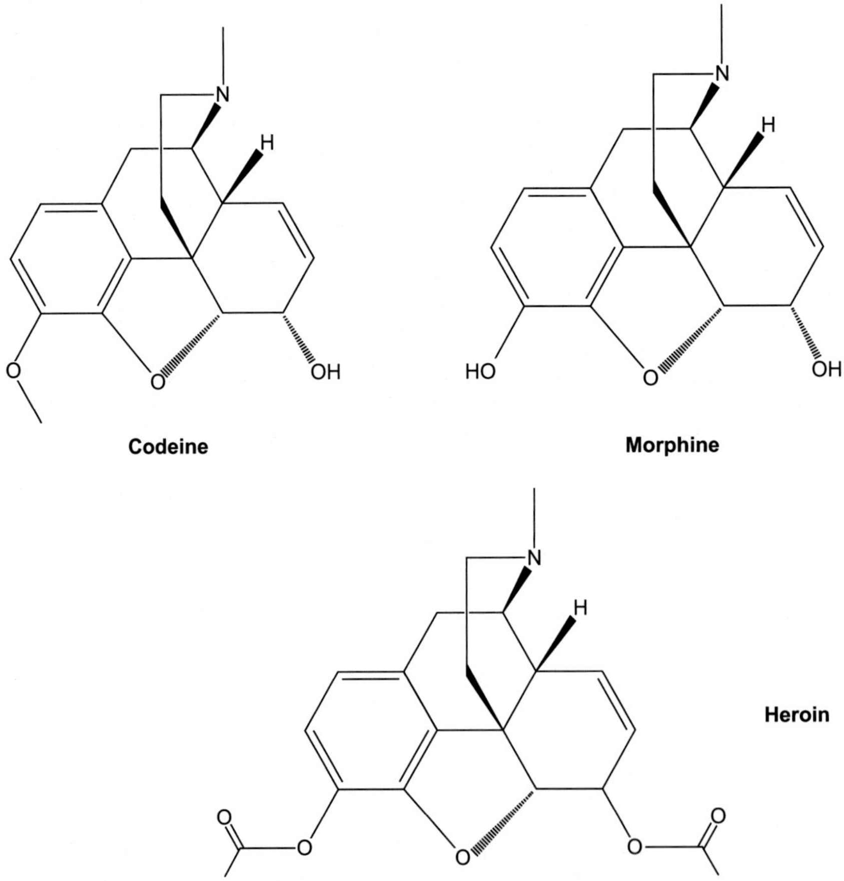

Codeine

Morphine

Heroin

FIG. 1.1 Chemical structure of codeine, morphine, and heroin.

CODEINE AND THEBAINE

Codeine (7,8-didehydro-4,5-epoxy-3-methoxy-17-methylmorphinan-6-ol monohydrate) is a naturally occurring constituent of the opium poppy, *P. somniferum*. It was isolated from opium in 1833, and its pain-relieving effects were recognized shortly after. Codeine constitutes about 0.5% of opium, which continues to be a useful source of its production, although the bulk of codeine used medicinally today is prepared by the methylation of morphine. Codeine is less potent than morphine, with a potency ratio of 1:10 [13]. Chemical structure of codeine is given in Fig. 1.1. Medical use of codeine is discussed in Chapter 2.

Thebaine, another alkaloid isolated from opium, is almost devoid of any analgesic effect. It is used as a starting material for synthesis of other opioids.

HEROIN

Heroin is the first designer drug that was synthesized by British pharmacist Alder Wright from morphine in 1874. Chemically, heroin is diacetylmorphine, which is acetylated form of morphine. The motivation of synthesis of heroin was to discover a novel compound with analgesic properties of morphine but without any abuse potential. He named the product "Heroin" because of its heroic properties as an analgesic [4]. Chemical structure of heroin is given in Fig. 1.1. Heroin was first marketed as an antitussive for patient with asthma and tuberculosis in 1898. However, after nearly 12 years of use, clinical rationale for using heroin medically was challenged because of its abuse potential. Because heroin is about as twice as potent as morphine as a cough suppressant, the small dose of heroin used for treating those patients may not have showed abuse potential initially, but later it was evident that heroin is highly addictive. The question of addiction became a serious issue in 1912 when reports showed that addicts used heroin because it was easily available over the counter in various formulations. In December 1914, US congress passed the Harrison Act that limited maximum amount of heroin in proprietary preparations less than 10 mg per gram of product. The total ban of heroin in the United States was achieved in 1924 [14].

Heroin is currently classified as a Schedule I drug with no known medical use but very high abuse potential. Street names of heroin are given in Table 1.3. Although heroin has been abused in over 100 years, during 2002–13, heroin overdose death rates nearly quadrupled in the United States, from 0.7 deaths to 2.7 deaths per 100,000 population, with a near doubling of the rates from 2011 to 13 [15]. Data from the National Survey on Drug Use and Health indicate

TABLE 1.3 Street Names of Opium and Heroin.	
Opium	Chinese molasses, Ope, black stuff
Heroin	Brown sugar, dope, junk, white horse, smack, H, tar, chiba, mud, etc.
Opium	Big O, black stuff, buddha, chillum, Chinese molasses, gum, hop, etc.

that heroin use, abuse, and dependence have increased in recent years. In 2013, an estimated 517,000 persons reported past-year heroin abuse or dependence, a nearly 150% increase since 2007 [16].

In recent years, a number of mainstream media reports that the abuse of heroin has migrated from low-income urban areas with large minority populations to more affluent suburban and rural areas with primarily white populations. Part of this increase in heroin abuse and apparent migration to a new class of users coincides with prescription opioid abuse over last 20 years. Cicero et al. reported, based on their review of survey results, that there are demographic differences between recent heroin abusers and people who started abusing heroin 40–50 years ago. It is interesting to note that recent users of heroin are older, white men, and women currently living primarily in nonurban areas who started taking opioids through valid prescription. Unfortunately, some of them suffered from opioid addiction and switched to cheaper and more accessible heroin over prescription opioid such as oxycodone (OxyContin). In contrast, earlier surveys indicated that in 1960 and 1970s, heroin abuse was an inner city issue among minority population. Therefore, with introduction to opioids for pain management, certain population (white middle-class men and women in less urban areas) were first exposed to opioid through prescription and later developed addiction. They may have chosen heroin because it is cheaper alternative to prescription opioids. Another interesting aspect of the study is that the age at first opioid use has increased over the past 50 years from 16 years of age to 23 years of age [17].

Other surveys have indicated that during 2002–11, rates of heroin initiation were reported to be highest among males, persons aged 18–25 years, non-Hispanic whites, those with an annual household income <$20,000, and those residing in the northeast. However, during this period, heroin initiation rates generally increased across most demographic subgroups. Most heroin users have a history of nonmedical use of prescription opioid pain relievers, and an increase in the rate of heroin overdose deaths has occurred concurrently with an epidemic of prescription opioid overdoses. However, when opioid prescriptions were restricted, heroin abuse was increased. Interestingly, analysis of 2010–12 drug overdose deaths in 28 states found that decreases in prescription opioid death rates within a state were not associated with increases in heroin death rates; in fact, increases in heroin overdose death rates were associated with increases in prescription opioid overdose death rates. In addition, a study examining trends in opioid pain reliever overdose hospitalizations and heroin overdose hospitalizations between 1993 and 2009 found that increases in opioid pain reliever hospitalizations predicted an increase in heroin overdose hospitalizations in subsequent year [18,19].

Heroin Synthesis and Cutting of Heroin

The majority of clandestine laboratories first isolate morphine from the crude opium extracts using "Lime Method." In this method, apolar impurities are removed by dissolving the opium in hot water followed by filtration, addition of calcium hydroxide, and again filtration. Then solution is boiled and ammonium chloride is added in order to precipitate morphine. Laboratories producing morphine as an end product perform further purification, but clandestine laboratories preparing illicit heroin may not carry any extra step to purify morphine. Therefore, heroin thus produced consists of a mixture of morphine and traces of other opium alkaloids, like codeine, noscapine, papaverine, and thebaine. In nearly all regions of the world, the "lime method" is used for further synthesis of heroin. The traditional "large-scale" synthesis, however, is a simple one-step acetylation reaction and is typically performed by the addition of a large excess of acetic anhydride directly to the crude morphine, followed by heating the resulting solution to, or near, boiling. Although acetic anhydride is by far the most common acetylating reagent used, on rare occasions, acetyl chloride and ethylidene diacetate may also be used as acetylating agent. After acetylation, reaction mixture is cooled and the intended product is isolated by treating the reaction mixture with sodium carbonate and collecting the heroin base by filtration. The resulting product is mostly used for smoking. For IV injection, the base heroin is further transformed into granular heroin hydrochloride by dissolving it in acetone and addition of hydrochloric acid until the solution changes color with blue litmus. The content is dried in air, and powdered heroin is produced by grinding. During acetylation, other alkaloids, found as impurities in the original morphine and having functional groups that can be acetylated, will produce several rearrangement and acetylation by-products such as acetylcodeine, acetylthebaol, or the very unstable 3-monoacetylmorphine. After heroin is formed, a small part will spontaneously deacetylate into 6-monoacetylmorphine, a major impurity of all heroin samples. Finally, cutting substances are added in order to increase the bulk of the product to increase net profit from selling illicit heroin. Sugars, such as glucose, lactose, and mannitol, are the most commonly used diluents. Often samples are cut with more than one diluent. In addition, also pharmacologically active substances or adulterants are incorporated into the cut heroin samples. Some of these components are added to enhance the effect of heroin. Acetaminophen, for example, increases the volatility of the heroin base and thus increases the effect by "chasing the dragon." Others, however, are added to facilitate the use. For example, addition of procaine is most likely done to locally relief the pain of an IV injection. Because of the large number of parameters that can vary during the preparation of heroin, it was thought that specific profiles, characteristic for the different heroin-producing countries or regions, could be constituted by comparative analysis [20].

Specific compounds used for adulteration of urine may vary depending on geographic locations. For example, in Malaysia, where the first report of heroin abuse was published in 1972, the first few clandestine heroin laboratories that synthesized heroin via the acetylation of imported morphine were uncovered in 1973 and 1977. Interestingly, by the mid-1980s, this type of laboratory was replaced by heroin-cutting laboratories, whereby imported high-grade heroin was cut to street heroin. This was to meet the rising demand for the drug owing to the rapid escalation of the number of drug users. Over the years, the most significant change in the composition of the street heroin is the decrease in its purity from 30%–50% to 3%–5%. Caffeine has remained the major adulterant, and chloroquine is detected in virtually all recent seized heroin products [21].

Concerning cutting of heroin in Europe, between 1960 and 1970s, substances such as caffeine, quinine, lactose, and mannitol were used. In the 1980s, caffeine, procaine, acetaminophen (paracetamol), and phenobarbital were common adulterants, while quinine was less frequently detected in seized heroin specimens. At the beginning of the 1990s, procaine, phenobarbital, and methaqualone disappeared as adulterants in heroin products abused in Europe. Currently, caffeine and acetaminophen (more than 90% specimens) are the major adulterants used for cutting heroin. In addition, glucose, sucrose (saccharose), lactose, mannitol, and even inositol have been found as diluents in illicit heroin specimens [22]. Schneider and Meys also reported that paracetamol (acetaminophen) and caffeine are the most commonly encountered adulterants in illicit heroin specimens in Europe [23]. Heroin adulterated with scopolamine may cause severe anticholinergic toxicity in heroin users with patients often presenting to emergency departments in great numbers [24]. Clenbuterol is a β2-adrenergic agonist with veterinary uses, which has not been approved by the US Food and Drug Administration for human use. Clenbuterol is infrequently reported as heroin adulterant, but if used, it may cause serious adverse reactions and may require hospitalization. Hieger et al. reported cases of ten patients presented with unexpected symptoms (chest pain, dyspnea, palpitation, and nausea/vomiting) shortly after heroin use. All patients were male, with a median age of 40 years (range: 38–46 years). The authors detected clenbuterol in all patients, thus explaining cause of toxicity. Fortunately, all patients survived with supportive care [25].

Risser et al. investigated potential link between purity of street heroin– and heroin-related deaths in Vienna in 1999. The authors analyzed a total of 415 seized heroin specimens with a total weight of 128.02 g. The median purity of heroin specimens was 6.5% (range: 0.0%–47% diacetylmorphine). All the samples contained a diluent, mainly lactose, as well as adulterants, such as caffeine and/or paracetamol (acetaminophen). During the study period, the authors investigated 75 heroin-related deaths and 387 heroin-related emergencies but did not observe any correlation between death related to heroin abuse and purity of street heroin. The authors concluded that the widely held belief that the number of heroin-related deaths could be explained simply through fluctuations in the purity of street heroin could not be substantiated, even though the results of this study do not rule out an association between the purity of heroin and heroin-related deaths/emergencies [26]. Toprak and Cetin concluded that the weight of heroin and the number of heroin seizures, but not the heroin purity, were significantly associated with the number of heroin-related deaths [27]. In contrast, in another study, the authors observed that the occurrence of overdose fatalities was moderately associated with both the average heroin purity and the range of heroin purity over the study period [28].

White versus Brown and Black Tar Heroin

Pure heroin is white in color but depending on origin heroin may be brown or look like black tar. Exclusive regional heroin markets have developed worldwide: poppies grown in Afghanistan predominately end up as heroin in Europe; Southeast Asian heroin almost exclusively goes to Australia and Western Canada; and Colombian and Mexican heroin enters the United States illegally. Prior to 1980, heroin in the United States was sourced from the three predominate producing regions in the world: Afghanistan, Pakistan, and Iran as well as the Golden Triangle of Southeast Asia (Myanmar, Laos, and Thailand) and to some extent Mexico. Opium from the Golden Triangle accounted for 55% of the world's illegal opium and the heroin derived from it accounted for 19% of the US market. The market share of these sources has changed dramatically over the last two decades, especially since a new source of heroin, Colombian, began exporting in the early 1990s. By 2000, the distribution of US retail heroin changed dramatically with 48% coming from Colombia, 39% from Mexico, 3% from Southwest Asia, 1% from Southeast Asia, and 9% unclassified. This trend has only continued, with the latest data showing that market share has increase to 58% for Colombian-sourced heroin, 40% for Mexican-sourced heroin, with Southeast and Southwest Asia heroin combined falling to 2% share. Heroin has several characteristics that affect how it is used and even how it is distributed. These characteristics include color, physical state, water solubility, pH balance, heat stability, and purity. Southeast Asian heroin is typically white, powdered, highly water soluble and acidic, while heroin coming from Southwest Asian is typically a brown coarse powder with poor water solubility (until acidified from its basic form by the addition of an acid) and good heat stability; Colombian heroin is off-white to light brown, powdered, and acidic with good water solubility. However, Mexican heroin is dark brown to black, solid, vaporizable, and of lower purity, and despite its acidity, it requires heat to dissolve this into aqueous solution. The risk of bacterial infections is well established with heroin use, especially use of black tar heroin [29].

In the United States, white and light brown powder heroin from Columbia predominate in East Coast drug markets, while Mexican black tar heroin predominates West Coast drug markets. Typical black tar heroin preparation includes stirring an injectable portion of black tar heroin and water over heat until the heroin dissolves. Powder heroin can be prepared for injection by stirring it with room temperature water. Heating heroin solution above 65°C inactivates HIV living in the heating vessel, which may partially explain why rates of HIV infection are lower in settings with black tar heroin, like California, compared to settings dominated by powder heroin, like the Eastern United States [30].

In the last decade, heroin source types in San Francisco have expanded beyond black tar heroin, with the addition of reportedly more potent gunpowder heroin that is a sticky powder version of black tar heroin. It may appear in a solid form that crumbles and other times as a mix of chunks and powder or with a "dried coffee" powdered appearance and sometimes speckled white and black. Gunpowder heroin has gained popularity as a higher potency alternative to the longer established product. Like black tar heroin, gunpowder heroin should be heated to dissolve quickly but tended to be more soluble than black tar heroin [31].

Heroin Laced with Fentanyl

In the United States, there is a dramatic increase in the adverse consequences of heroin use as number of deaths related to heroin overdose is increasing yearly. This is due to transitioning a portion of the at-risk population from one source of opioids (prescription pills) to a fully illicit drug (heroin). In addition, contamination of the heroin supply with powerful synthetic opioids, especially fentanyl and its chemical analogues, results in a public health crisis associated with opioid-related deaths. Fentanyl is the central chemical in a family of over 20 analogues; it is a synthetic opioid with potent μ-receptor activity, registering 30−50 times more potent by weight compared with heroin. Therefore, when heroin is cut with more potent fentanyl, the users have no idea that they are abusing fentanyl. This cutting of heroin with fentanyl significantly increases heroin-related overdoses as well as fatalities [32]. Use of fentanyl to cut heroin is to increase the profit margin because heroin costs $65,000 per kilogram while fentanyl is available roughly at $3500 per kilogram. Producing precise fentanyl dosage demands special equipment and knowledge, but street suppliers of heroin who cut heroin with fentanyl have neither the knowledge nor the required equipment. Therefore, risk of heroin overdose is substantially higher in addicted people who abuse heroin cut with fentanyl [33].

A public health emergency was declared in British Columbia, Canada, in April 2016 due to the unprecedented number of deaths from opioid overdoses, where fentanyl abuse was a major reason. While in 2012, fentanyl was implicated in 5% of drug overdose deaths; the proportion has increased to nearly 80% in 2017. Amlani et al. reported that about one-third of clients tested positive for fentanyl. Among those who tested

positive, 73% were not aware of their fentanyl exposure [34]. Based on a study of 24 participants, Mema et al. concluded that fentanyl urine testing appealed to illicit opioid users and may have contributed to adopting behaviors toward safer drug use. A relationship of trust between tester and client seemed important for clients who expressed concerns with privacy of the urine test results [35].

Baldwin et al. reported that the numbers of seized fentanyl and heroin samples, as well as both total illicit drug overdose deaths and fentanyl-detected overdose deaths, are increasing rapidly in British Columbia since 2009. The authors identified a positive association between seized fentanyl and total unintentional overdose deaths that is stronger than the relationship between seized heroin and total unintentional overdose deaths [36]. Number of seized heroin and seized heroin cut with fentanyl between 2008 and 2016 in British Columbia, Canada, are given in Table 1.4.

Heroin Pharmacokinetics

During heroin synthesis from morphine, initial acetylation occurs at position 3 of the phenolic hydroxyl group of morphine. Then the phenolic hydroxyl group at position 6 is also acetylated, producing diacetylmorphine (heroin). The chemical addition of the ester groups to morphine renders heroin more lipophilic than morphine. Therefore, heroin can cross the blood-brain barrier much faster than morphine explaining early onset of pharmacological effects compared to

TABLE 1.4
Number of Seized Drugs Containing Heroin or Heroin Laced With Fentanyl Between 2008−16 in British Columbia, Canada.

Year	Heroin	Heroin Plus Fentanyl	Percent Laced with Fentanyl
2008	1021	0	0%
2009	748	0	0%
2010	674	1	0.15%
2011	960	0	0%
2012	1139	3	0.26%
2013	1318	49	3.7%
2014	1608	67	4.2%
2015	2072	262	12.6%
2016	2178	1349	61.9%

Source of data: Reference Amlani A, McKee G, Khamis N, Raghukumar G et al. Why the FUSS (Fentanyl Urine Screen Study)? A cross-sectional survey to characterize an emerging threat to people who use drugs in British Columbia, Canada. Harm Reduct J. 2015;12:54.

morphine. However, opiate receptors in the brain are stereospecific. As a result, heroin has lower affinities for opiate receptors than morphine. However, heroin metabolite such as 6-monoacetylmorphine (also known as 6-acetylmorphine [6-AM]) has much higher affinity for opiate receptors. In addition, 6-monoacetylmorphine is also metabolized into morphine. Therefore, heroin can be considered as a "prodrug," where pharmacological effects are mostly due to its active metabolites, 6-monoacetylmorphine and morphine [37].

The ionization constant of heroin (pKa) is 7.6. Therefore, at physiological pH, average 40% heroin will be in nonionizable state and accessible for membrane transport. In contrast, pKa of morphine is 9.4. The protein binding is 20%—40%. After entering into circulation, heroin is rapidly hydrolyzed into 6-monoacetylmorphine and finally into morphine. The hydrolysis of heroin and also 6-monoacetylmorphine is catalyzed by different types of esterase enzymes: acetylcholinesterase in erythrocytes and pseudocholinesterase in plasma. Then morphine is conjugated to produce morphine-3-glucuronide (major metabolite) and also morphine-6-glucuronide (minor active metabolite). Glucuronidation is catalyzed by the liver enzyme uridine-5′-diphosphate glucuronosyltransferase. Morphine glucuronides are hydrophilic compounds and are mostly excreted in urine, but a minor fraction is also excreted in bile. After intravenous administration of heroin, approximately 70% of dosage is recovered in urine mainly as morphine glucuronide [38].

Because of rapid hydrolysis by serum esterases, heroin blood levels decline rapidly after intravenous administration and become undetectable after 10—40 min. The half-life is 1.3—7.8 min. Unchanged heroin is not usually detected in urine. The maximum concentration of 6-monoacetylmorphine, the first hydrolysis product of heroin in blood, is achieved rapidly (0.7—2.7 min) after intravenous administration of heroin. However, half-life (5.2—52 min) is much longer than half-life of heroin. After heroin injection, 6-monoacetylmorphine could be detected in serum 1—3 h postadministration. This metabolite is also detected in urine. However, 6-monoacetylmorphine is further hydrolyzed into morphine, and maximum concentration could be achieved 3.6—8.0 min after administration of heroin. The half-life of morphine as a heroin metabolite is 100—280 min in blood [38].

The pharmacodynamic effect of heroin depends on the pharmacokinetic profile of heroin and is dependent on route of administration. The immediate effect of intravenously injected heroin that is often described by heroin addicts as a "flash" (warm and pleasant sensation). The intensity of flash is related to maximum serum levels of heroin and 6-monoacetylmorphine. The flash is followed by euphoria, which is related to morphine and morphine-6-glucuronide plasma levels. Because heroin is lipophilic in nature, it is also rapidly absorbed after intranasal application as heroin is rapidly absorbed through the mucus membrane. Maximal heroin concentration in serum could be achieved in 2—5 min after intranasal application. However, peak heroin concentrations are significantly lower after intranasal application than intravenous injection, hence less pharmacological effect. Heroin, like other opioids, causes respiratory distress and concomitant use of alcohol and increases the risk of heroin overdose [38].

For workplace drug testing as well as medical drug testing, the presence of 6-monoacetylmorphine in urine at a concentration of 10 ng/mL is considered as the confirmation of heroin abuse. Immunoassays are commercially available with a cutoff concentration of 10 ng/mL for detecting 6-monoacetylmorphine in urine [39]. In general, 6-monoacetylmorphine can be detected in serum up to 2 h and in urine up to 8 h after heroin abuse. This is the major limitation of using 6-monoacetylmorphine as the marker for heroin abuse. However, detection of 6-monoacetylmorphine in blood confirms recent exposure to heroin. In addition, 6-monoacetylmorphine in vitreous humor is more stable because esterases are not present in vitreous. Therefore, measuring 6-monoacetylmorphine in vitreous humor is useful in death investigations where heroin overdose is suspected [40].

Adverse Effects of Heroin

Heroin addiction may change normal brain structure and function. Chronic heroin abuse results in a long-lasting impairment in cognitive function and alteration in the normal functions of central nervous system. Studied have shown that chronic heroin abuse has adverse effects on prefrontal cortex, temporal insula, thalamus, nucleus accumbens, amygdala, and sensomotor structure of the brain. Investigation using magnetic resonance imaging of the brain of heroin addicts indicated that heroin abuse is associated with decreased gray matter density in the prefrontal and temporal cortical regions of the brain. Because gray matter is associated with muscle control, sensory perception, memory, emotion, speech, and decision-making abilities of the brain, reduced gray matter in the brain of a chronic heroin addicts is associated with impaired brain functions. Electrophysiological studies of the brain functions of heroin addicts using electroencephalography have shown major deviations compared to normal subjects. The cognitive dysfunction of heroin abusers is correlated with $\alpha 2$ mean frequency shift at the central region. The duration of heroin addiction is correlated with frequency shift of $\alpha 2$ region, and heroin dosage is associated with lower $\alpha 1$ mean frequency at central, temporal, and occipital sites [41]. Shen et al. reported diminished cortical plasticity in heroin-addicted subjects that indicates the involvement of the motor system in advanced stages of addiction. One potential mechanism underlying the diminished cortical plasticity is the loss of dopaminergic signaling at the synaptic site, which reduces induction of plasticity [42].

Heroin use is associated with blood disseminated bacterial infections, e.g., endocarditis and osteomyelitis. More recently, association between abuse of black tar heroin and bacterial skin infection as well as soft tissue infections has been reported. Black tar heroin has been associated with wound botulism, necrotizing fasciitis, tetanus, and soft tissue infections. These infections may be due to contamination of the heroin or by function of subcutaneous injection in users with heroin-scarred veins. Clostridium species, e.g., those bacteria that cause botulism and tetanus, may be able to survive the harsh chemical environment of black tar heroin by forming spores; indeed, these spores may be made more viable by the heating required to solubilize tar heroin.

Infective endocarditis is a life-threatening condition with a high mortality rate. Intravenous drug abusers are at high risk of developing infective endocarditis. Although most of the cases of infective endocarditis are caused by a single pathogen, cases of polymicrobial endocarditis are rare and they are associated with a reported mortality rate of more than 30%. Mehrzad et al. reported a case of a 49-year-old male with intravenous heroin and fentanyl abuse, who presented with infective endocarditis caused by *Neisseria sicca/subflava*, *Actinomyces*, *Streptococcus mitis*, and *Haemophilus parainfluenzae*, complicated by septic emboli to the lungs and skin, splenic infarct, and immunocomplex-mediated proliferative glomerulonephritis. The patient survived the infection [43].

Botulism is a rare, treatable yet potentially fatal disorder caused by toxins produced by some bacteria that affects the presynaptic synaptic membrane resulting in a characteristic neuromuscular dysfunction. Most common cause of botulism is from eating improperly canned food and honey by infants. More recently, there has been an increased incidence of wound botulism associated with intramuscular and subcutaneous injection of street drugs, especially heroin in particular

[44]. Wound botulism can be deadly, and it carries a case fatality rate of 15%, mostly secondary to ventilatory failure [45].

Following the introduction of black tar heroin mainly from Mexico in the 1980s, cases of wound botulism dramatically increased in the Western United States. Contamination with spores of *Clostridium botulinum* of black tar heroin occurs along the distribution line. The heating of heroin powder to solubilize it for subcutaneous injection ("skin popping") does not kill the spores. The spores germinate in an anaerobic tissue environment and release botulinum toxin type A or B. Skin abscesses on the skin in a suspected drug abuser may help in diagnosis. However, definite diagnosis is made by detection of botulinum toxin in serum or isolation of *C. botulinum* from the abscess. Early treatment with equine ABE botulinum antitoxin obtained from the Centers for Disease Control and Prevention is useful for good treatment outcome [46]. Qureshi et al. reported cases of 15 patients with mean age of 47 years (12 men and 3 women) who suffered from wound botulism due to abuse of black tar heroin. All patients had abscesses in the administration areas. By history, the most common symptoms were dysphagia (66%), proximal muscle weakness of upper and lower extremity (60%), neck flexor muscle weakness (33%), ophthalmoplegia (53%), bilateral ptosis (46%), dysarthria (53%), double vision (40%), blurred vision (33%), and dry mouth (20%). In patients with documented wound botulism, the pupils were reactive in 46%. All patients required mechanical ventilation and were treated with the trivalent antitoxin. Eleven patients (73.3%) were discharged home, two were transferred to a skill nursing facility, and two were transferred to long-term acute care facility [47]. Wound botulism due to abuse of IV drugs including black tar heroin may also be associated with fatality [48].

Heroin addiction is associated with poor oral health. Oral and dental care in heroin addiction might be complicated by altered mental status, negative attitude toward oral health, dental anxiety and fear, drug interaction, and associated medical comorbidity [49].

BLOOD AND URINE LEVELS OF HEROIN AND ITS METABOLITES

Heroin can be analyzed by gas chromatography/mass spectrometry (GC/MS) without derivatization, but morphine requires derivatization prior to GC/MS analysis. Trimethylsilyl derivative is a commonly used derivative, but many other derivatization methods have been reported. Guillot et al. reported an analytical method for the determination of heroin, free 6-AM, and free morphine in blood, urine, and vitreous humor for analysis of postmortem cases where heroin was suspected to be the cause of overdose death. The authors synthesized diethylnalorphine from nalorphine to use as the internal standard. After liquid-liquid extraction at pH to 9.5, the authors converted 6-monoacetylmorphine into 6-acetyl-3-propanoylmorphine and morphine into dipropanoylmorphine prior to GC/MS analysis. The propionylation step was conducted at room using propionic anhydride, while 4-dimethylaminopyridine was also used as a catalyst. Neither heroin nor diethylnalorphine (internal standard) is transformed during this derivatization. The reaction products are analyzed using full-scan (250–405 amu) ion trap GC/MS. This method provides a baseline separation and distinctive mass spectrum of each compound of interest [50]. Heroin and its metabolites can also be analyzed using high-performance liquid chromatography or liquid chromatography combined with mass spectrometry (LC-MS).

Death due to heroin abused is widely reported in the literature. Analysis of heroin, 6-monoacetylmorphine, and morphine in biological fluids is very useful in both medical and forensic investigations. The metabolism of heroin offers a potential approach whether death due to heroin overdose is rapid or delayed. After administration, heroin is deacetylated into 6-monoacetylmorphine within an average time of 3 min. The average conversion time of 6-monoacetylmorphine into morphine is 22 min. Therefore, the presence of 6-monoacetylmorphine in the blood is suggestive of survival times of less than 20–30 min following heroin administration. Darke and Duflou investigated 145 cases of death (mean age: 40.5 years, 81% were male) related to heroin abuse by measuring concentrations of 6-monoacetylmorphine, free morphine, morphine-3-glucuronide, and morphine-6-glucuronide in blood. The authors detected the presence of 6-monoacetylmorphine in 43% of cases. The median-free morphine concentration in 6-monoacetylmorphine–positive cases was more than twice that of cases without the presence of 6-monoacetylmorphine (0.26 vs. 0.12 µg/mL). The authors concluded that in heroin-related deaths in Sydney, Australia, during 2013 and 2014, 6-monoacetylmorphine was present in the blood in less than half of cases, suggesting that a minority of cases had survival times after overdose of less than 20–30 min. Therefore, detection of 6-monoacetylmorphine in blood indicates rapid death after heroin overdose [51].

Rop et al. reported a case of a 40-year-old man who had a long history of heroin abuse. He was placed in a detention center and was under medical supervision. Every evening before being locked up, he was searched and in one evening during search attempted to hide some drug by swallowing. He told the guard that it was a capsule of bromazepam. He was found dead in his cell on the next morning. Toxicology analysis showed the presence of heroin (109 ng/mL), 6-monocaetylmorphine (168 ng/mL), and morphine (1140 ng/mL) in his blood, but in urine, only morphine was detected at a very high concentration (3650 ng/mL). The authors also showed the presence of heroin (17 ng/gm), 6-monoacetylmorphine (12 ng/gm), and morphine (425 ng/gm) in the gastrointestinal content. However, no bromazepam was detected during toxicological analysis. The authors concluded that the person died due to heroin overdose because under clinical condition and after constant infusion of heroin hydrochloride (20 mg per 180 min), blood concentrations of steady state of heroin, 6-monoacetylmorphine, and morphine were reported at 57, 15, and 30 ng/mL. Heroin and 6-monoacetylmorphine could only be detected in blood after 5 min, but morphine in blood could be detected up to 45 min after controlled administration of heroin. However, in the deceased, heroin, 6-monocaetylmorphine, and morphine could be detected in blood for a longer time, indicating massive overdose. In addition, heroin, 6-monoacetylmorphine, and morphine concentrations in the blood of deceased were 2-, 11-, and 38-folds higher than those reported in the steady state. The moment of death was probably at the time when peak morphine concentration was achieved, which was during the night [52].

Hair analysis of heroin, 6-monoacetylmorohine, and morphine is very useful in demonstrating long-term heroin abuse. A 5-year-old child in a comatose state was admitted to the hospital with respiratory arrest, scarce cardiac activity, hypothermia, cyanosis, miosis, and total unconsciousness. He responded to therapy with naloxone. Urine drug screen using immunoassay was positive for opiates and barbiturates. When the child regained consciousness, he reported that one of his relatives often forced him to consume bitter and unpleasant brown and white powder. The judge put the child in custody of a social assistant after he was released from the hospital and also ordered toxicological analysis of child's hair. Scalp hair grows at a rate of 1.5 cm/month, and when drugs are consumed on a regular basis, they accumulate in the hair matrix in the zone of the shaft relative to that period of growth. Segmental hair analysis was performed in order to determine if drug consumption had continued. Quantitative results showed the presence of morphine and 6-monoacetylmorphine in increasing amounts from the second to the distal segment of hair, each approximately corresponding to a period of time of 1 month. The 6-monoacetylmorphine concentrations ranged from 0.2 ng/mg of hair in the second segment of hair to 0.6 ng/mg of hair in the distal segment. The morphine concentrations ranged from 0.1 to 0.3 ng/mg of hair. The proximal segment of hair, which corresponded to the period that the child was housed in a social center, was negative for opiates. The judge accepted results of toxicology analysis. The relative of the child was found guilty [53].

Tassoni et al. reported a case of a 39-year-old man who was a known heroin addict for last 2 years. He arrived in hospital in a coma caused by a cardiocirculatory failure due to heroin overdose, as supported by a witness statement and confirmed by the hospital toxicological analysis. After 5 days, the man died and death was attributed to cardiac arrest. The quantitative results of the GC/MS analysis performed 144 h after admission to the hospital showed a high morphine level of 21.3 μg/mL in the bile, but no morphine was detected in the blood. However, morphine was also detected in his hair at a concentration of 4.8 ng/mg of hair. Usually morphine can be detected in blood and urine up to 2 days after heroin abuse. The authors concluded that bile analysis is a viable option to demonstrate heroin overdose when morphine, the metabolite of heroin, cannot be detected in blood or urine [54].

Codeine is often present in heroin preparations as an impurity and is not a metabolite of heroin. Studies report that a ratio of morphine to codeine greater than one indicates heroin use. Ellis et al. investigated a total of 166 cases with quantifiable morphine and codeine in postmortem blood specimens. From these cases, 163 deaths had morphine/codeine ratio of greater than 1 and 108 cases had 6-monoacetylmorphine detected in a postmortem sample. The authors found no statistically significant difference between individuals with a history of intravenous drug abuse and individuals with no known history of intravenous drug abuse with respect to either morphine/codeine ratio >1 or the presence of 6-monoacetylmorphine. The authors concluded that morphine/codeine ratio >1 in an IV drug user is sufficient evidence to conclude heroin use by a decedent even if 6-monoacetylmorphine is not detected in a postmortem sample [55].

Alunni-Perret et al. reported a case of death by heroin overdose in an embalmed body. The deceased, a 30-year-old French white male, was found dead in Thailand several days after a scuffle; the precise circumstances of his death were unknown. However, the deceased was a known drug addict. Before repatriation to France, the body was embalmed. External examination of the embalmed body did not show any signs of violence. Also there was no evidence indicating drug injection. In addition, there was no evidence of natural disease and no anatomical cause of death was found. All the internal organs were heavy, saturated with formaldehyde. The toxicological study was performed in the bile and in the liver, using GC/MS analysis. No alcohol was found in the bile, but high levels of morphine (2476 ng/mL) and codeine (305 ng/mL) were found in the bile. In addition, analysis of liver tissue showed a morphine level of 4.3 mg/kg and hair analysis showed the presence of 6-monoacetylmorphine (6.99 ng/mg of hair), morphine (4.21 ng/mg of hair), and codeine (0.23 ng/mg of hair). No other drug was detected. The authors concluded that the cause of death was heroin overdose and commented that as usual in cases of embalmment, fluids such as blood and urine were unavailable, and toxicological analysis could still be performed using bile and the liver because several drug can still be detected in fluids and tissues that contain formaldehyde due to embalmment of corpse. This case demonstrates that in embalmed corpses, toxicological assessment is still possible, e.g., after heroin fatalities [56].

Treatment of Heroin Overdose and Heroin Withdrawal

Naloxone is the antidote for patients suffering from opioid-induced respiratory depression or arrest including heroin overdose. Because naloxone has a short half-life, usually patients are observed for 4−6 h because of concern for rebound toxicity. Given this concern, paramedics often transport patients who receive naloxone for medical observation for several hours. In addition after awakening, some heroin users attempt to refuse further treatment or transport to the hospital. This causes considerable problem for paramedics and ED staff caring for the patient as they weigh attempts to act in the patient's best interest against holding someone without consent. Concern and controversy exist about whether heroin users require observation after naloxone, how long after the last dose of naloxone one must be observed, and the risks—including recurrence of respiratory depression or occurrence of pulmonary edema—incurred by these patients. Willman et al. based on their study commented that the risk of death in overdosed patients successfully treated with naloxone is very small. Therefore, heroin abusers who responded to naloxone therapy and have normal vital signs may be released, provided they are monitored by friends [57]. The initial dose of naloxone is 0.4—2 mg IV or alternatively may be administered by intramuscular injection or subcutaneously. If response is not observed, doses may be repeated 2—3 min interval. In cases of heroin overdose, if respiratory depression is not reversed following repeated intravenous bolus doses of naloxone—a total of 10—12 mg of naloxone can be given as intravenous boluses—then it is not recommended to set up an intravenous infusion [58].

Several types of medications have been used for stabilizing heroin users: methadone, buprenorphine, natrexone, and levo-α-acetylmethadol. The pharmacotherapies used to treat opioid dependence have also been linked to opioid poisoning. The long-acting opioid agonist, methadone, has been associated with high rates of opioid poisoning in the first 2 to 4 weeks following initiation of treatment and also first 2 weeks after discontinuation of treatment. Similarly, the opioid antagonist naltrexone has been associated with high rates of opioid poisoning mortality, following the cessation of treatment, due to a reduction of opioid tolerance and a rapid unblocking of mu opioid receptors after discontinuation of naltrexone. Kelty and Hulse studied opioid-dependent patients treated with methadone (n = 3515), buprenorphine (n = 3250), or implant naltrexone (n = 1461) in Western Australia between 2001 and 2010 and concluded that rates of fatal and nonfatal opioid overdose were not significantly different in patients treated with methadone, buprenorphine, or implant naltrexone [59].

The prescription of heroin for the management of heroin dependence is a controversial treatment approach that was limited to Britain until the 1990s. Since then, a number of countries such as Switzerland, Holland, and Germany have embarked upon clinical trials of this approach, and it is currently licensed and available in several European countries. Heroin is usually prescribed in intravenous dosages of 300—500 mg/day, divided in two or three doses. Uncommon but serious side effects include seizures and respiratory depression immediately following injection. Heroin treatment results in a comparable retention, improved general health and psychosocial functioning, and less self-reported illicit heroin use than oral methadone treatment. Cost-effectiveness studies indicate heroin treatment to be more expensive to deliver but may offer savings in the criminal justice sector. There has been debate regarding how heroin treatment should be positioned within the range of treatment approaches for this condition. There is increasing consensus that, in countries that have robust and accessible treatment systems for heroin users, heroin treatment is suited to a minority of heroin users as a second-line treatment for those individuals who do not respond to methadone or buprenorphine treatment delivered under optimal conditions [60]. Therefore, prescription heroin administration should be the last resort for treating heroin addicts when other therapies are ineffective. Moreover, heroin therapy should only be used in clinical settings where proper follow-up is ensured [61].

CONCLUSIONS

Opium use is known from ancient time and morphine is the first narcotic analgesic known to mankind. However, abuse of opium was widespread in ancient time, and now abuse of morphine and especially heroin is a public health and safety issue. Although effective treatment using naloxone is available for reversing life-threatening opiate overdose including heroin overdose, death from heroin abuse is widespread. Effective intervention and treatment is essential to fight opioid abuse crisis including heroin abuse crisis.

ACKNOWLEDGMENTS

All structures presented in this chapter are drawn by Mathew D. Krasowski, MD, PhD, Vice Chair of Clinical Pathology, University of Iowa, Roy J., and Lucille A., Caver College of Medicine, Iowa City, IA. Figures are courtesy of Dr. Krasowski.

REFERENCES

[1] Schmidt J, Boettcher C, Kuhnt C, Zenk MH. Poppy alkaloid profiling by electrospray tandem mass spectrometry and electrospray FT-ICR mass spectrometry after [ring-13C]-tyramine feeding. Phytochemistry 2007;68:189—202.

[2] Kłys M, Rojek S, Maciów-Głąb M, Kula K. Opium alkaloids in toxicological medico-legal practice of department of forensic medicine, jagiellonian university medical college. Arch Med Sadowej Kryminol 2013;63:301—6 [Article in Polish].

[3] Norn S, Kruse PR, Kruse E. History of opium poppy and morphine. Dan Medicinhist Arbog 2005;33:171—84 [Article in Danish].

[4] Khademi H, Kamangar F, Brennan P, Malekzadeh R. Opioid therapy and its side effects: a review. Arch Iran Med 2016;19:870—6.

[5] Stefano GB, Pilonis N, Ptacek R, Kream RM. Reciprocal evaluation of opiate science from Medical and cultural perspective. Med Sci Mon Int Med J Exp Clin Res 2017;23:2890—6.

[6] Obladen M. Lethal lullabies: a history of opium use in infants. J Hum Lactation 2016;32:75—85.

[7] Moosavyzadeh A, Ghaffari F, Mosavat SH, Zargaran A, et al. The medieval Persian manuscript of Afyunieh: the

first individual treatise on the opium and addiction in history. J Integr Med 2018;16:77—783.

[8] Fernandez H, Libby LA. Heroin: its history, pharmacology, and treatment. (The Library of Addictive Drugs) Hazelden Publishing; February 1, 2011.

[9] Liu C, Hua Z, Bai Y. Classification of opium by UPLC-Q-TOF analysis of principle and minor alkaloids. J Forensic Sci 2016;61:1615—21.

[10] Berridge V. Heroin prescription and history. N Engl J Med 2009;361:820—1.

[11] Remberg B, Sterrantino AF, Artner R, Janitsch C, Krenn L. Science in drug control: the alkaloid content of afghan opium. Chem Biodivers 2008;5:1770—9.

[12] Narayanaswami K, Golani HC, Dua RD. Assay of major and minor constituents of opium samples and studies of their origin. Forensic Sci Int 1979;14:181—90.

[13] Williams DG, Hatch DJ, Howard RF. Codeine phosphate in paediatric medicine. Br J Anaesth 2001;86:413—21.

[14] Sneader W. The discovery of heroin. Lancet 1998;352: 1697—9.

[15] Hedegaard H, Chen LH, Warner M. Drug-poisoning deaths involving heroin: United States, 2000—2013. NCHS Data Brief 2015;(190):1—8.

[16] Substance Abuse and Mental Health Services Administration. Results from the 2013 national survey on drug use and health: summary of national findings. Rockville, MD: US Department of Health and Human Services, Substance Abuse and Mental Health Services Administration; 2014. NSDUH Series H-48, HHS Publication No. (SMA) 14-4863.

[17] Cicero TJ, Ellis MS, Surratt HL, Kurtz SP. The changing face of heroin use in the United States: a retrospective analysis of the past 50 years. JAMA Psychiatry 2014;71: 821—6.

[18] Unick GJ, Rosenblum D, Mars S, Ciccarone D. Intertwined epidemics: national demographic trends in hospitalizations for heroin- and opioid-related overdoses, 1993—2009. PLoS One 2013;8(2):e54496.

[19] Bauman ZM, Morizio K, Singer M, Hood CR, et al. The heroin epidemic in America: a surgeon's perspective. Surg Infect 2019 [E-pub ahead of print].

[20] Dams R, Benijts T, Lambert WE, Massart DL, De Leenheer AP. Heroin impurity profiling: trends throughout a decade of experimenting. Forensic Sci Int 2001;123:81—8.

[21] Sulaiman M, Kunalan V, Yap ATW, Lim WJL, et al. Heroin in Malaysia and Singapore. Drug Test Anal 2018;10: 109—19.

[22] Broséus J, Gentile N, Esseiva P. The cutting of cocaine and heroin: a critical review. Forensic Sci Int 2016;262:73—83.

[23] Schneider S, Meys F. Analysis of illicit cocaine and heroin samples seized in Luxembourg from 2005—2010. Forensic Sci Int 2011;212:242—6.

[24] Hamilton RJ, Perrone J, Hoffman R, Henretig FM, et al. A descriptive study of an epidemic of poisoning caused by heroin adulterated with scopolamine. J Toxicol Clin Toxicol 2000;38:597—608.

[25] Hieger MA, Emswiler MP, Maskell KF, Sentz JT, et al. A case series of clenbuterol toxicity caused by adulterated heroin. J Emerg Med 2016;51:259—61.

[26] Risser D, Uhl A, Oberndorfer F, Hönigschnabl S, et al. Is there a relationship between street heroin purity and drug-related emergencies and/or drug-related deaths? An analysis from Vienna, Austria. J Forensic Sci 2007; 52:1171—6.

[27] Toprak S1, Cetin I. Heroin overdose deaths and heroin purity between 1990 and 2000 in Istanbul, Turkey. J Forensic Sci 2009;54:1185—8.

[28] Darke S, Hall W, Weatherburn D, Lind B. Fluctuations in heroin purity and the incidence of fatal heroin overdose. Drug Alcohol Depend 1999;54:155—61.

[29] Ciccarone D. Heroin in brown, black and white: structural factors and medical consequences in the US heroin market. Int J Drug Pol 2009;20:277—82.

[30] Ciccarone D, Bourgois P. Explaining the geographical variation of HIV among injection drug users in the United States. Subst Use Misuse 2003;38:2049—63.

[31] Mars SG, Bourgois P, Karandinos G, Montero F, Daniel Ciccarone D. Textures of heroin: user perspectives on "black tar" and powder heroin in two US cities. J Psychoact Drugs 2016;48:270—8.

[32] Ciccarone D, Ondocsin J, Mars SG. Heroin uncertainties: exploring users' perceptions of fentanyl-adulterated and -substituted 'heroin'. Int J Drug Pol 2017;46:146—55.

[33] Frank RG, Pollack HA. Addressing the fentanyl threat to public health. N Engl J Med 2017;376:605—7.

[34] Amlani A, McKee G, Khamis N, Raghukumar G, et al. Why the FUSS (Fentanyl Urine Screen Study)? A cross-sectional survey to characterize an emerging threat to people who use drugs in British Columbia, Canada. Harm Reduct J 2015;12:54.

[35] Mema SC, Sage C, Popoff S, Bridgeman J, et al. Expanding harm reduction to include fentanyl urine testing: results from a pilot in rural British Columbia. Harm Reduct J 2018;15:19.

[36] Baldwin N, Gray R, Goel A, Wood E, et al. Fentanyl and heroin contained in seized illicit drugs and overdose-related deaths in British Columbia, Canada: an observational analysis. Drug Alcohol Depend 2018;185:322—7.

[37] Inturrisi CE, Schultz M, Shin S, Umans JG, et al. Evidence from opiate binding studies that heroin acts through its metabolites. Life Sci 1983;33(Suppl. 1):773—6.

[38] Rook EJ, Huitema AD, van den Brink W, van Ree JM, Beijnen JH. Pharmacokinetics and pharmacokinetic variability of heroin and its metabolites: review of the literature. Curr Clin Pharmacol 2006;1:109—18.

[39] Holler JM, Bosy TZ, Klette KL, Wiegand R, et al. Comparison of the Microgenics CEDIA heroin metabolite (6-AM) and the Roche Abuscreen ONLINE opiate immunoassays for the detection of heroin use in forensic urine samples. J Anal Toxicol 2004;28:489—93.

[40] Dinis-Oliveira RJ. Metabolism and metabolomics of opiates: a long way of forensic implications to unravel. J Forensic Leg Med 2019;61:128—40.

[41] Motlagh F, Ibrahim F, Menke JM, Rashid R, et al. Neuro-electrophysiological approaches in heroin addiction research: a review of literatures. J Neurosci Res 2016;94: 297—309.

[42] Shen Y, Cao X, Shan C, Dai W, Yuan TF. Heroin addiction impairs human cortical plasticity. Biol Psychiatr 2017;81: e49—50.

[43] Mehrzad R, Sublette M, Barza M. Polymicrobial endocarditis in intravenous heroin and fentanyl abuse. J Clin Diagn Res 2013;7:2981—5.

[44] Abavare L, Abavare C. Wound botulism resulting from heroin abuse: can you recognize it? J Emerg Nurs 2012; 38:301—3.

[45] Shapiro RL, Hatheway C, Swerdlow DL. Botulism in the United States: a clinical and epidemiologic review. Ann Intern Med 1998;129:221—8.

[46] Davis LE, King MK. Wound botulism from heroin skin popping. Curr Neurol Neurosci Rep 2008;8:462—8.

[47] Qureshi IA, Qureshi MA, Rauf Afzal M, Maud A, et al. Black tar heroin skin popping as a cause of wound botulism. Neurocrit Care 2017;27:415—9.

[48] Peak CM, Rosen H, Kamali A, Poe A, et al. Wound botulism outbreak among persons who use black tar heroin — San diego county, California, 2017—2018. Morb Mortal Wkly Rep 2019;67(5152):1415—8.

[49] Abed H, Hassona Y. Oral healthcare management in heroin and methadone users. Br Dent J 2019;226:563—7.

[50] Guillot JG, Lefebvre M, Weber JP. Determination of heroin, 6-acetylmorphine, and morphine in biological fluids using their propionyl derivatives with ion trap GC-MS. J Anal Toxicol 1997;21:127—33.

[51] Darke S, Duflou J. The toxicology of heroin-related death: estimating survival times. Addiction 2016;111:1607—13.

[52] Rop PP, Fornaris M, Salmon T, Burle J, Bresson M. Concentrations of heroin, 06-monoacetylmorphine, and morphine in a lethal case following an oral heroin overdose. J Anal Toxicol 1997;21:232—5.

[53] Strano Rossi S, Offidani C, Chiarotti M. Application of hair analysis to document coercive heroin administration to a child. J Anal Toxicol 1998;22:75—7.

[54] Tassoni G, Cacaci C, Zampi M, Froldi R. Bile analysis in heroin overdose. J Forensic Sci 2007;52:1405—7.

[55] Ellis AD, McGwin G, Davis GG, Dye DW. Identifying cases of heroin toxicity where 6-acetylmorphine (6-AM) is not detected by toxicological analyses. Forensic Sci Med Pathol 2016;12:243—7.

[56] Alunni-Perret V, Kintz P, Ludes B, Ohayon P, Quatrehomme G. Determination of heroin after embalmment. Forensic Sci Int 2003;134:36—9.

[57] Willman MW, Liss DB, Schwarz ES, Mullins ME. Do heroin overdose patients require observation after receiving naloxone? Clin Toxicol 2017;55:81—7.

[58] Vahabzadeh M, Banagozar Mohammadi A. Heroin body-packing and naloxone. Lancet 2019;393(10177):e35.

[59] Kelty E, Hulse G. Fatal and non-fatal opioid overdose in opioid dependent patients treated with methadone, buprenorphine or implant naltrexone. Int J Drug Pol 2017;46:54—60.

[60] Lintzeris N. Prescription of heroin for the management of heroin dependence: current status. CNS Drugs 2009;23: 463—76.

[61] Ferri M, Davoli M, Perucci CA. Heroin maintenance for chronic heroin-dependent individuals. Cochrane Database Syst Rev 2011;12:CD003410.

CHAPTER 2

Prescription Opioids: An Overview

INTRODUCTION

Pain is the most common reason a person seeks medical attention. It has been estimated that over 100 million Americans are living with chronic pain [1]. Acute pain is different than chronic pain because peripheral neurons (nociceptors) are capable of detecting painful stimulus such as extreme change of temperature, pressure, tissue damage, or by chemicals most often related to inflammation. Acute pain resolves when external stimulus no longer exists or when inflammation or tissue damage is resolved. However, management of chronic pain is medically challenging.

There are known demographic factors that predispose a person to develop chronic pain. Johannes et al. based on a survey of 27,035 individuals in the United States reported that weighted prevalence of chronic pain (defined as chronic, recurrent, or long-lasting pain lasting for at least 6 months) was 30.7% among the population surveyed. Interestingly, prevalence of chronic pain was higher for females (34.3%) than males (26.7%) and increased with age. The weighted prevalence of primary chronic lower back pain was 8.1% and primary osteoarthritis pain was 3.9%. Half of respondents with chronic pain experienced daily pain, and average (past 3 months) pain intensity was severe (≥ 7 on a scale ranging from 0 to 10) for 32% of individuals surveyed by the authors. Multiple logistic regression analysis showed low household income and unemployment as significant socioeconomic indicators of chronic pain [2]. Shmagel et al. also reported that individuals with lower annual household income have greater odds of reporting chronic pain compared with persons with higher annual income [3]. Moreover, chronic pain and psychiatric disorders frequently cooccur. McWilliams et al. based on a survey of 5877 individuals representing United States population observed significant positive associations between chronic pain and individual 12-month mood and anxiety disorders. The strongest associations were observed with panic disorder and posttraumatic stress disorder [4]. The total healthcare costs for chronic pain treatment are estimated to range between $560 to 635 billion per year in the United States, which are more than the annual costs of heart disease, diabetes, and cancer [5].

HISTORY OF PAIN MANAGEMENT

Pain is the oldest medical problem. European physicians used opium for pain relief for a long time, and after 1680, laudanum, the mixture of opium in sherry introduced by Thomas Sydenham, was widely prescribed for pain management. William T.G. Morton performed his famous demonstration of anesthesia with ether in 1846. The British obstetrician James Young proposed the use of chloroform in child birth and surgery in 1848. However, throughout the 19th century, opiates were the standard therapy for both acute and chronic pain. Morphine was industrially produced in Germany in the 1820s and was produced in the United States a decade later. Then with the discovery of hypodermic syringe in 1855 by Alexander Wood, pain management using morphine was improved because such syringe made administration of morphine so convenient that it probably also contributed to morphine abuse. During that time, opium and alcohol-based compounds in the form of liquid, pill, or headache powder were freely available over the counter from pharmacies and many people self-medicated them with these opium-containing products. By the 1870s, physicians were concerned about morphine abuse. In 1898, Bayer introduced diacetylmorphine under the trade name "heroin" as a cough suppressant. However, by 1910, young American working class people started abusing heroin, and finally, heroin became banned and now is a Schedule I drug with no known medical benefit but having a very high abuse potential (see also Chapter 1). Bayer's chemist not only synthesized heroin but also synthesized acetyl salicylate (aspirin), which turned out to be remarkably safe and effective analgesic. Bayer aspirin became an over-the-counter drug in 1917. In 1965, the Canadian psychologist Ronald Melzack and British physiologist Patrick Wall, building on the hypothesis of Dutch surgeon Willem Noordenbos, published their classic "gate control" paper proposing a spinal cord mechanism that regulates the transmission of pain sensation between the periphery and the brain. Then, John Bonica in 1973 formed an interdisciplinary organization known as "The International Association for the Study of Pain." Research in past 30 years resulted

Fighting the Opioid Epidemic. https://doi.org/10.1016/B978-0-12-820075-9.00002-8

in development of a variety of alternatives to opiate pain management [6]. Nevertheless, opioids are widely used for pain management today. This chapter focuses on opioids used for pain management, as well as opioids used for reversal of opioid overdoses.

OPIOID RECEPTORS

All opioids interact with opioid receptors present in the brain for their pharmacological activities. Wide interindividual variations in response to opioids are due to genetic polymorphisms of genes encoding opioid receptors, as well as liver enzymes metabolizing opioids, for example, CYP2D6 isoenzyme. This topic is discussed in detail in Chapter 4. In this section, a brief overview of opioid receptors is presented.

Opioid receptors are expressed by both central and peripheral neurons, as well as by neuroendocrine, immune, and ectodermal cells. There are three main types of opioid receptors (μ, δ, and κ receptors). Additional opioid receptor types such as sigma, epsilon, and orphanin have been proposed, but currently, they are no longer considered as "classical opioid receptors." The opioid receptors belong to the class A gamma subgroup of seven transmembrane G protein—coupled receptors and demonstrate 50%—70% homology between their genes [7].

The μ-opioid receptor is most important for pain management because it plays central role in analgesic activity of morphine and most opioids. Morphine, hydrocodone, oxycodone, hydromorphone, fentanyl, etc., are strong agonist of μ-opioid receptor, while codeine, tramadol, etc., are weak agonist. Morphine is also a weak agonist of both δ- and κ-opioid receptors. Buprenorphine is a partial agonist of μ-opioid receptor, while it is antagonist of κ receptor. Naloxone and naltrexone are antagonists of all three opioid receptors and can reverse effects of opioid. In addition, some endogenous opioid peptides are also agonist of opioid receptors. There are three distinct families of opioid peptides: endorphins, encephalin, and dynorphins. Endorphins mostly bind to μ-opioid receptors. Various features of opioid receptors are summarized in Table 2.1.

PRESCRIPTION OPIOIDS

Morphine was the first prescription opiate used for treating pain. Since then, many more opioids have been marketed and approved by the FDA for pain management. Most recently, in October 2018, the Food and Drug Administration (FDA) approved the opioid analgesic Dsuvia, which is a sublingual formulation of Sufentanil, for use in "medically supervised health care settings." Dsuvia is up to 10 times more potent than fentanyl [8].

TABLE 2.1
Various Features of Opioid Receptors.

Opioid Receptor	Comments	Agonist	Antagonist	Location
μ receptor	μ receptor was the first receptor characterized, and it is the major opioid receptor responsible for pain control because morphine, codeine, and synthetic opioids bind with this receptor. Binding of drugs with μ receptors is also responsible for adverse effects.	Morphine, codeine synthetic opioids Endorphins[a]	Naloxone Naltrexone (strongest interaction)	Brain Spinal cord Submucosal plexus Mesenteric plexus
δ receptor	Responsible for analgesia, respiratory depression, and physical dependance.	Enkephalins[a]	Naloxone Naltrexone Buprenorphine	Brain Mesenteric plexus
κ receptor	Responsible for analgesia and sedation but contribute little to addiction.	Pentazocine Dynorphins[a]	Naloxone Naltrexone Buprenorphine	Brain Spinal cord Mesenteric plexus

[a] Endogenous opioid peptides.

Both opiates and opioids are used in pain management. Opiate is referred to naturally occurring alkaloids found in opium such as morphine and codeine and semisynthetic alkaloids with similar structures such as buprenorphine, dihydrocodeine, heroin, hydrocodone, hydromorphone, oxycodone and oxymorphone. The term opioids refer to compounds that are structurally different from natural opiates but are narcotic analgesic because they interact with opioid receptors similar to natural opiates. Opioids can be subclassified under three broad categories: natural alkaloids from opium, semisynthetic opioids, and synthetic opioids (Table 2.2). Opioids can also be classified by chemical structure (Table 2.3). Based on interactions with opioid receptors, opioids used in pain management can be classified as strong agonist, moderate agonist, weak agonist, mixed agonists (interacting with multiple opioid receptors), and mixed opioid agonists/antagonists. Drugs such as naloxone and naltrexone, which are used for reversing opioid overdose, are antagonists for opioid receptors (Table 2.4). Commonly used opioids in pain management are listed in Table 2.5. Several opioids have active metabolites, and some of these active metabolites are also used as individual drugs. Opioids that have active metabolites are listed Table 2.6. All opioids are small molecules. Their molecular weights are listed in Table 2.7.

Starting in the mid-1990s, allegations arose that the medical field systematically undertreated pain, and the American Pain Society lobbied to have pain recognized as a fifth vital sign that, if adopted, would require all physicians to accept and treat patient complaining about pain. As a result, there was an increase in opioid prescriptions and increasing profits for drug

TABLE 2.2
Commonly Used Natural, Semisynthetic, and Synthetic Opioids.

Natural Alkaloids From Opium	Semisynthetic Opioids	Synthetic Opioids
Morphine	Hydromorphone	Fentanyl
Codeine	Hydrocodone	Alfentanil
	Oxymorphone	Sufentanil
	Oxycodone	Remifentanil
	Buprenorphine	Meperidine
	Dihydrocodeine	Methadone
	Dihydromorphine	Levorphanol
	Ethylmorphine	Pentazocine
		Tramadol
		Tapentadol

TABLE 2.3
Classification of Opioids Based on Chemical Structure.

Chemical Structure	Individual Drugs
Phenanthrene	Morphine, codeine, buprenorphine, hydrocodone, hydromorphone, levorphanol, oxycodone, oxymorphone, naloxone, nalbuphine
Benzomorphan	Pentazocine, loperamide
Phenylpiperidines	Fentanyl, alfentanil, sufentanil, remifentanil, meperidine
Diphenylheptane	Methadone, propoxyphene
Phenylpropylamine	Tramadol, tapentadol

manufacturers [9]. At the same time, published papers in medical journals suggested that cancer patients using prescription opioids did not become addicted. In fact, Schug et al. reported that only 1 patient out of 550 developed an addiction to their prescription painkillers [10]. Another study reported no cases of addiction among 10,000 burn victims using prescription opioid drugs [11]. Such studies convinced clinicians that opioids could be used safely for pain management [12]. By 2000, the Joint Commission began requiring that healthcare organizations assess and treat pain in all patients. OxyContin prescriptions for non–cancer-related pain increased from 670,000 in 1997 to nearly 6.2 million in 2002 [13]. It has been estimated that in 2012, approximately 259 million opioid prescriptions had been written—enough for every adult in America to have at least one bottle of pills [9].

Opioid Prescription Pattern
While opioid prescribing has been decreasing in the United States since 2013, the amount of opioids prescribed is still three times higher than in 1999. In 2017, over 35% of opioid overdose deaths involved a prescription opioid [14]. Currently, opioids are one of the most commonly prescribed drugs in the United States. Although the rate of opioid prescribing is high, a large portion of all opioid prescriptions comes from a small group of prescribers and varies considerably by prescriber specialty. In one study, the authors reported that between July 1, 2016, and June 30, 2017, a total of 209.5 million opioid prescriptions were dispensed in the United States. Primary care physicians

TABLE 2.4
Classification of Opioid Drugs Based on Interactions With Opioid Receptors.

Type of Interaction	Examples of Individual Drugs	Comments
Strong μ-opioid receptor agonists	Heroin, morphine, fentanyl, fentanyl derivatives, methadone	Heroin has no know medical use while morphine, fentanyl, and fentanyl derivatives are used in pain management, but these drugs are also addictive. Methadone used for drug rehabilitation is also addictive
Moderate μ-opioid receptor agonists	Codeine, hydrocodone, and oxycodone	Oxycodone and hydrocodone are also addictive
Weak μ-opioid receptor agonist	Tramadol	Use for managing moderate pain
Mixed opioid receptor agonists	Pentazocine	Partial agonist of μ receptor and agonist of κ receptor
Mixed opioid receptor agonists/antagonists	Buprenorphine	Partial agonist for μ receptor but antagonist for κ receptor and weak antagonist for δ receptor
Antagonists for opioid receptors	Naloxone, naltrexone	Antagonists of all three opioid receptors but major actions are through interactions with μ receptors. Naloxone is better for treating opioid overdose because it is short acting.

TABLE 2.5
Commonly Prescribed Opioids.

Opioid	Pharmacokinetic Parameters
Morphine	Oral bioavailability: 23.8% Volume of distribution: 2.1−4 L/kg. Plasma protein binding: approximately 35%. Metabolites: morphine-3-glucuronide and morphine-6-glucuronide that are excreted in urine. Minor metabolite: normorphine
Codeine	Oral bioavailability: approximately 90% Volume of distribution: 3−6 L/kg Plasma protein binding: 22%−29% Major metabolite is codeine-6-glucuronide, but morphine, a metabolite of codeine, may account for some analgesic activities. Minor metabolites include free and conjugated hydrocodone, hydromorphone, norhydrocodone, norhydromorphone, normorphine.
Oxycodone	Oral bioavailability is 60%−80%, volume of distribution is 2−4 L/kg, and serum protein binding is approximately 45%. Peak serum concentration is observed 1−2 h after ingesting capsule but 0.5−1.5 h after administration of oral solution. Oxycodone is metabolized by both CYP3A and CYP2D6. Only 9% of dose is recovered in urine as unchanged oxycodone, while major metabolites are noroxycodone and noroxymorphone.
Hydrocodone	Maximum serum level observed within 1 h or oral administration with eliminating half-life of 4−6. However, for extended release (once a day formulation), maximum serum concentration may be observed after 6−30 h. The average volume of distribution is 5.7 L/kg. Major metabolites are norhydrocodone and hydromorphone. Hydromorphone is an active metabolite responsible for pain control.

(continued)

TABLE 2.5 Commonly Prescribed Opioids.—cont'd	
Opioid	**Pharmacokinetic Parameters**
Hydromorphone	Hydromorphone has advantage over morphine in pain management. Oral bioavailability: 50% Volume of distribution: 1.22 L/kg Half-life is 2—3 h but extended release formulation has much longer half-life and can be administered once daily. Major urinary metabolite is hydromorphone-3-glucuronide.
Oxymorphone	Oxymorphone is a μ-opioid agonist which is more active than morphine. Oral bioavailability is approximately 10%. Protein binding: 10%—12% Elimination half-life is 7—10 h for immediate release formulation but 9—12 h for extended release formulation. Major urinary metabolite is oxymorphone-3-glucuronide, while <2% drug is excreted unchanged in urine.
Fentanyl	Fentanyl is 80—100 times more potent than morphine. After intravenous administration, effects are observed within minutes. Oral bioavailability: approximately 50%. Fentanyl is 80%—86% bound to serum proteins. Volume of distribution: 3—8 L/kg Overlap between therapeutic and toxic concentrations. Major metabolite: norfentanyl
Methadone	Methadone is a synthetic opioid that is used for treating patients with opioid addiction (methadone maintenance program) and also for pain management. Oral bioavailability: approximately 75% (upper end up to 100%) Serum protein binding: 71%—88%, mostly to α-1-acid glycoprotein, protein binding to serum albumin minimal. Mean elimination half-life of R-methadone is 37.5 h, which is longer than mean elimination half-life of S-methadone (28.6 h). Volume of distribution is higher for R-methadone than S-methadone. Major inactive metabolites of methadone are 2-ethyl-1,5-dimethyl-3,3-diphenylpyrrolidine (EDDP) and 2-ethyl-5-methyl-3,3-diphenylpyrroline (EMDP).
Meperidine	Meperidine is no longer considered as the first choice opioid drug because other opioids with better safety profiles are available. After oral administration only 50%—60% of dose reaches the systematic circulation. The volume of distribution is approximately 3.5 L/kg. Meperidine is approximately 65%—75% bound to serum proteins. Meperidine is metabolized by the liver where hydrolysis of meperidine produces meperidinic acid, and N-demethylation results in the formation of normeperidine. Neurotoxicity of meperidine is mostly due to its metabolite normeperidine.
Tramadol	Tramadol is a racemic mixture of R (+) and S (−) tramadol. Tramadol is an atypical opioid because in addition to being a weak agonist of μ-opioid receptor, it also modulates the monoaminergic system by inhibiting noradrenergic and serotoninergic reuptake. Tramadol is available for oral and parenteral administration. Oral formulation may be immediate release or extended release. Oral bioavailability is high and serum protein binding is approximately 20%. Tramadol is mainly metabolized by O- and N-demethylation and by conjugation reactions forming glucuronides and sulfates. Tramadol and its metabolites are mainly excreted via the kidneys. The mean elimination half-life is about 6 h.
Tapentadol	Tapentadol is a dual action drug which acts as μ-opioid receptor agonist and also as norepinephrine reuptake inhibitors

(continued)

TABLE 2.5
Commonly Prescribed Opioids.—cont'd

Opioid	Pharmacokinetic Parameters
	Oral absorption is approximately 32%. Half-life is 4 h and during action is 6 h. Very few drug interaction because Phase 1 metabolism is a minor pathway, while conjugation with glucuronic acid and sulfate represent major metabolic pathway.
Buprenorphine	Buprenorphine is used for treating patients addicted to opioids, as well as in pain management. Buprenorphine is a preferred opioid for treatment of pain in patients with renal or liver dysfunction. Buprenorphine bioavailability is 49% after administration of sublingual solution and 29% with sublingual tablets. Buprenorphine is approximately 96% protein bound to α- and β-globulins. Buprenorphine is extensively metabolized in the liver, and the major metabolite is norbuprenorphine. Both buprenorphine and norbuprenorphine are further conjugated with glucuronic acid during Phase II metabolism. Buprenorphine is eliminated primarily via a stool, while 10%—30% of the dose is excreted in urine as conjugated forms of buprenorphine and norbuprenorphine.
Levorphanol	Levorphanol is a unique synthetic opioid due to its wide range of activities, including agonist for μ-opioid receptor, as well as both δ- and κ-opioid receptors. In addition, levorphanol is an antagonist of N-methyl-D-aspartate receptor, as well as reuptake inhibitor of both norepinephrine and serotonin. The terminal half-life is 11—16 h, and duration of analgesic effect is 6—15 h. Major metabolite is levorphanol-3-glucuronide.
Propoxyphene	Withdrawn from US market in 2010.

TABLE 2.6
List of Commonly Used Opioids Which Have Active Metabolite.

Opioid Drug	Active Metabolite	Inactive Metabolite	Active Metabolite Which is Also a Drug
Morphine	Morphine-6-glucuronide Hydromorphone (minor metabolite)	Morphine-3-glucuronide Normorphine	Hydromorphone
Codeine	Morphine, Hydrocodone (minor metabolite)	Norcodeine	Morphine Hydrocodone
Hydrocodone	Hydromorphone	Norhydrocodone	Hydromorphone
Oxycodone	Oxymorphone	Noroxycodone	Oxymorphone

(family medicine, internal medicine, general practice) accounted for 37.1% of all prescriptions; nonphysician prescribers (physician assistant, nurse practitioner) accounted for 19.2%; and pain medicine specialists accounted for 8.9%. Compared with previous research, the results suggest decreases in the share among primary care physicians and increases among nonphysician prescribers and pain medicine specialists [15].

Nataraj et al. reported that among all opioid prescribers, the five most common specialties were internal medicine (23.8%), family medicine (15%), surgery (10.5%), emergency medicine (6.7%), and obstetrics/gynecology (5.7%). The most common specialties among the sampled high-volume prescribers were family medicine (32.2%), internal medicine (22.9%), orthopedics (11.4%), emergency medicine (6.8%), and pain medicine (5.8%). The median age for high-volume prescribers in the sample was 53 years (compared with 50 years for all prescribers). Overall, as well as across all specialties, hydrocodone had the highest mean number of prescriptions, followed by oxycodone, and tramadol was the third most frequently prescribed opioid type across most categories of specialists, with a few exceptions. These exceptions include prescription of morphine by specialists of pain medicine, radiology, oncology, and

TABLE 2.7
Molecular Weight of Common Opioids.

Opioid	Molecular Weight
Morphine	285.3
Codeine	299.3
Hydrocodone	299.3
Oxycodone	315.3
Hydromorphone	285.3
Oxymorphone	301.3
Fentanyl	335.5
Alfentanil	416.5
Sufentanil	386.5
Remifentanil	376.4
Buprenorphine	467.7
Tramadol	263.4
Tapentadol hydrochloride	257.8
Meperidine	247.3
Methadone	309.4
Levorphanol	257.4
Naloxone	327.4
Naltrexone	341.4

palliative medicine. In general, psychiatrists prescribed methadone, and dentists and obstetricians/gynecologists prescribed codeine more frequently than tramadol [16].

MORPHINE

Morphine is a strong narcotic analgesic that can be administered orally, intravenously, or by intramuscular injection. Diluted oral morphine solution and diluted tincture of opium are used for neonatal abstinence syndrome and for treating some adult conditions. Morphine is indicated for treating severe pain only where alternative treatment may be inadequate. Morphine is a strong agonist of μ-opioid receptor and also has a high abuse potential. After oral administration, morphine is absorbed with a median time to maximum blood concentration of 0.75 h. The oral bioavailability is 23.8%. Morphine is distributed mainly in extracellular water with a volume of distribution of 2.1—4 L/kg. The plasma protein binding is approximately 35%. Morphine is extensively metabolized by Phase II metabolism enzyme UDP-glucuronosyltransferase 2B7 (UGT2B7) into morphine-3-glucuronide (MG3, inactive metabolite)

and morphine-6-glucuronide (MG6, active metabolite). Minor metabolism via CYP enzymes such as CYP2C19 and CYP3A4 results in formation of normorphine. Another minor metabolite is hydromorphone. Excretion of MG3 and MG6 is predominately determined by renal function and is correlated with creatinine clearance [17,18]. Chemical structure of morphine is given in Chapter 1.

CODEINE

Codeine is an opiate analgesic and antitussive that is administered alone or in combination of another drug. Acetaminophen combined with codeine (Tylenol 2) is a commonly prescribed drug. Codeine is used in the treatment of mild-to-moderate pain. Although codeine may be used to treat chronic pain in cancer patients, its use to treat other types of chronic pain remains controversial. Chemical structure of codeine is given in Chapter 1.

Codeine has an oral bioavailability of approximately 90%, which is significantly higher than morphine. It has been postulated that codeine may exert its moderate analgesic potency through partial biotransformation to morphine by O-demethylation. Approximately 10% of codeine is converted into morphine, but this percent varies widely depending on CYP2D6 enzymatic activity. For Caucasian population, most people (77%—92%) are extensive wild-type metabolizers (normal metabolism of codeine), 2%—11% are intermediate metabolizers (less activity of CYP2D6 isoenzyme), 5%—10% are poor metabolizers (PMs, almost no enzymatic activity), and 1%—2% are ultrarapid metabolizers with high CYP2D6 activity due to gene duplication or other reason [19]. Therefore, polymorphism of CYP2D6 gene plays an important role in therapy with codeine. Please see Chapter 4 for more detail. Codeine is also excreted in urine after conjugation with glucuronic acid.

Trace amount of metabolites of codeine include free and conjugated hydrocodone, hydromorphone, norhydrocodone, norhydromorphone, normorphine, and 6α-hydrocodol, as well as 6β-hydrocodol [20]. Oyler et al. commented that analyses of the codeine formulations administered to subjects in their study revealed no hydrocodone present at the limit of detection of the assay (10 ng/mL). Therefore, hydrocodone can be produced as a minor metabolite of codeine in humans and may be excreted in urine at concentrations as high as 11% of parent drug concentration. Consequently, the detection of minor amounts of hydrocodone in urine containing high concentrations of codeine should not be interpreted as evidence of hydrocodone abuse [21].

West et al. based on their study with detection of codeine and its metabolites in oral fluid reported that only 15 oral fluid specimens (1.7%) out of 868 oral fluid specimens collected from patients taking codeine contained codeine, as well as codeine metabolites, such as morphine, hydrocodone, and norhydrocodone. However, 380 specimens (43.7%) were positive for codeine and any of its metabolites tested (morphine, hydrocodone, or norhydrocodone). In addition, 26 specimens (3.0%) were negative for codeine and morphine but positive for hydrocodone and norhydrocodone. In addition, 33 specimens (3.8%) tested positive for hydrocodone only. The authors conclude that hydrocodone is an important metabolite of codeine in the oral fluid [20].

Depending on polymorphism of CYP2D6, individuals could be PM (poor CYP2D6 enzyme activity) or extensive metabolizer (EM) with normal CYP2D6 activity. Yue et al. based on a study using 14 healthy Caucasians, including 8 EM and 6 PM of debrisoquine, observed no difference in the mean area under the curve (AUC), half-life, and total plasma clearance of codeine between extensive and PMs. The average half-life of codeine was 2.58 h (range: 1.55−3.29 h) and maximum serum level of codeine was observed approximately 1 h after oral administration of codeine in EMs and 0.86 h in PMs. However, PMs show significantly lower AUC of metabolites of codeine and morphine, MG3, MG6, and normorphine, compared with EMs. It is important to mention that in addition to morphine which is an active metabolite of codeine, MG6 and normorphine also have pharmacological activities. However, among EMs, the AUC of codeine-6-glucuronide was 15 times higher than that of codeine, which in turn was 50 times higher than that of morphine. The total recovery of drug-related material in 48 h urine collections ranged from 71% to 106% of the dose and did not differ between extensive and PMs. Six percent of the dose was O-demethylated in EMs and the majority of the metabolites produced through this pathway were conjugated. The unconjugated morphine accounted for less than 0.2% of the dose. In contrast in PMs, only 0.33% of the dose was metabolized by O-demethylation and morphine accounted for only 0.001% of the administered codeine. However, PMs showed higher concentration of norcodeine compared with EMs [22]. Interestingly, Chinese are less able to metabolize codeine compared to Caucasians mainly due to lower efficiency in glucuronidation [23].

Shah and Mason studied pharmacokinetics of codeine in 10 health volunteer after administration of 60 mg codeine sulfate. The mean peak codeine plasma concentrations was 88.1 ng/mL, which was achieved 1.2 h after oral administration of codeine sulfate. Mean maximum concentrations of metabolically produced morphine was 2.7 ng/mL. The mean ratio of areas under the plasma concentration-time curves for morphine and codeine was 0.027. Thus, free morphine represented only about $2.7 \pm 1.8\%$ of the free codeine area in each case [24]. The protein binding of codeine is 22%−29% and volume of distribution is 3−6 L/kg [25]. In another study, the fraction of codeine bound to plasma proteins, determined by ultrafiltration, was significantly higher in sickle cell patients ($66.0\% \pm 8.6\%$) than in healthy controls ($30.5\% \pm 2.7\%$). In general gamma globulin levels are increased in sickle cell patients. Codeine is known to bind to gamma globulin, a fact that may explain in part the observed increase in the plasma protein binding of codeine in sickle cell patients [26].

Chen et al. also studied the pharmacokinetics, metabolism, and partial clearances of codeine to morphine, norcodeine, and codeine-6-glucuronide after single (30 mg) and chronic (30 mg 8 h for seven doses) administration of codeine in eight subjects (seven extensive and one PM of dextromethorphan). After the single dose, maximum plasma codeine concentration was achieved at an average of 0.97 h, while maximum concentration of major metabolite codeine-6-glucuronide was observed at an average of 1.28 h. The plasma AUC of codeine-6-glucuronide was approximately 15.8 times higher than that of codeine. The AUC of codeine in saliva was on average 3.4 times higher than that in plasma. The average elimination half-life of codeine was 3.2 h and that of codeine-6-glucuronide was 3.2. The renal clearance of codeine was $183 + 59$ mL/min and was inversely correlated with urine pH. The renal clearance of codeine-6-glucuronide was $55 + 21$ mL/min and was not correlated with urine pH. After the single dose, approximately 86.1%, the dose was recovered in urine where codeine-6-glucuronide (average 59.8% of the recovered dose) represents the major metabolite of codeine, while total morphine (average 7.1% of the recovered dose) is the minor metabolite. In addition, 11.8% of dose was excreted in urine unchanged codeine and 6.9% was recovered in urine as norcodeine. These recoveries were not significantly different after chronic administration compared with single dose administration [27].

Oxycodone

Oxycodone was first synthesized from thebaine, a naturally occurring alkaloid in opium in 1917. Today, in addition to oxycodone, thebaine is also the starting material for synthesis of other drugs, including oxymorphone, naloxone, and buprenorphine. Oxycodone is a commonly prescribed opioid with high abuse potential (see Chapter 3). Similar to morphine, pharmacological activity of oxycodone is due to its interaction with μ-opioid receptor (agonist). However, its affinity for μ-opioid receptor is significantly lower compared to morphine. Interestingly, oxycodone metabolite, noroxymorphone, has two- to threefold higher affinities for μ-opioid receptor than oxycodone. Oxycodone is well absorbed after oral administration with a bioavailability of 60%−80%. Oral bioavailability of oxycodone is significantly higher than morphine. Oxycodone is approximately 40% bound to serum proteins, and volume of distribution is 2−5 L/kg in adults. The typical elimination half-life of oxycodone is 3−6 h [28]. Chemical structure of oxycodone is shown in Fig. 2.1.

Oxycodone metabolism is complex. Only 9% of administered dose is excreted in urine as oxycodone, mostly in unconjugated form. Urinary metabolites derived from CYP3A-mediated N-demethylation of oxycodone produces noroxycodone, accounting for approximately 23% administered dose. Noroxycodone is further metabolized by CYP2D6 into noroxymorphone (14% of administered dosage). In addition, noroxycodone is also converted into α- and β-noroxycodol via ketoreduction. These pathways account for approximately 45% of administered dosage. The CYP2D6-mediated O-demethylation produces oxymorphone (11% of administered dosage), α-, and β-oxymorphol, and 6-keto-reduction produces α- and β-oxycodol, which are other oxycodone metabolites found in urine after administration of oxycodone. In general, noroxycodone and noroxymorphone are the major metabolites in circulation with elimination half-lives longer than that of oxycodone. CYP3A-mediated N-demethylation is the principal metabolic pathway of oxycodone in humans. The central opioid effects of oxycodone are governed by the parent drug, with a negligible contribution from its circulating oxidative and reductive metabolites [29]. Nevertheless, polymorphisms of CYP2D6 affect therapy with oxycodone because oxymorphone has a strong analgesic effect and also is an opioid drug by itself [30].

Oxycodone is sometimes prescribed to pregnant women. In pregnant or laboring women, oxycodone passes freely through the placenta. As a result, oxycodone concentrations are similar in maternal and fetal plasma. Low concentrations of oxycodone are also detected in breast milk. Oxycodone interacts with many drugs. Concomitant use of CYP3A inhibitors, such as clarithromycin, itraconazole, ketoconazole, miconazole, voriconazole, and ritonavir, may increase serum oxycodone level due to reduced clearance. In contrast, CYP3A4

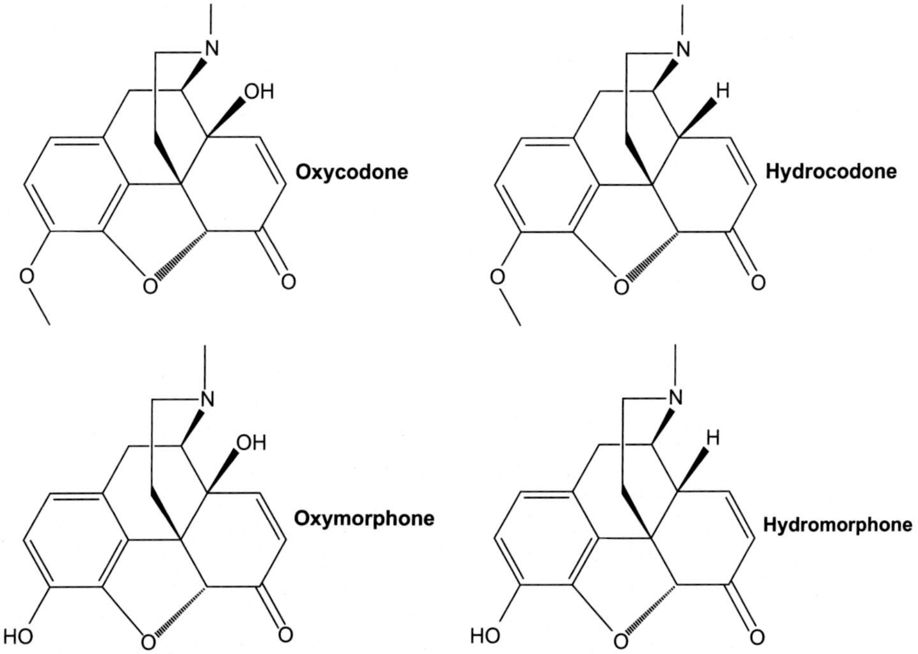

FIG. 2.1 Chemical structures of oxycodone, hydrocodone, oxymorphone, and hydromorphone.

inducers such as rifampin and St John's wort may increase clearance of oxycodone producing low plasma level and lack of inadequate pain control. Liver isoenzyme CYP2D6 also plays an important role in oxycodone metabolism. As expected, coadministration of CYP2D6 inhibitors, for example, quinidine and paroxetine, affects oxycodone metabolism, but such effects are less than CYP3A inducers and inhibitors. However, concomitant use of CYP2D6 inhibitors with CYP3A4 inhibitors may cause more pronounced effects of oxycodone. The interaction is more significant when oxycodone is administered orally, indicating some first-pass metabolism in the intestine [31].

HYDROCODONE

United States is the primary consumer of hydrocodone, using 99% of the global supply for 4.4% of the global population. Hydrocodone is a μ-opioid receptor agonist. The FDA has rescheduled hydrocodone from Schedule III to Schedule II due to its high abuse potential, which went into effect on October 6, 2014, along with a limit on added acetaminophen of 325 mg for each dose of hydrocodone [32]. Hydrocodone bitartrate extended-release (Hysingla ER) was the first single-entity hydrocodone formulation recognized by the FDA as having abuse-deterrent properties. Once-daily oral hydrocodone ER provides consistent plasma hydrocodone concentrations and sustained analgesia over the 24-h dosing interval. Its physicochemical properties render hydrocodone ER harder to manipulate physically, which is expected to deter intranasal, intravenous, and oral abuse. Hydrocodone ER is well tolerated, with a safety profile consistent with other narcotic analgesics, but, like other opioids, it is associated with risks of addiction, abuse/misuse, and serious adverse events, including respiratory depression, withdrawal, physical dependence, and overdose [33]. Chemical structure of hydrocodone is shown in Fig. 2.1.

Hydrocodone is available in combination with acetaminophen or ibuprofen (hydrocodone bitartrate 2.5—7.5 mg ibuprofen 200 mg). In February 2018, the FDA approved the combination of prodrug benzhydrocodone and acetaminophen for the short-term management of acute pain. After oral administration of immediate-release (IR) hydrocodone formulation, maximum serum concentration is achieved within 1 h with the elimination half-life of 4—6 h. However, following single oral dose of hydrocodone extended release (ER) formulations; peak serum concentration is observed at a median time of 14—16 h (range 6—30 h). The average volume of distribution is 5.7 L/kg. The major metabolites of hydrocodone are norhydrocodone and hydromorphone. Pain relief correlates with plasma hydromorphone but not with hydrocodone concentration, thus confirming that the ability to convert hydrocodone to its active drug is essential. Hydrocodone is metabolized to hydromorphone through O-demethylation catalyzed by the enzyme CYP2D6. Hydromorphone also undergoes Phase II glucuronidation to be transformed into metabolite hydromorphone-3-glucuronide. Polymorphisms of CYP2D6 play a role in pharmacological effects of hydrocodone. Approximately 3% of blacks and 1% of Asians are PMs of CYP2D6. These patients may not get adequate pain relief when treated with hydromorphone. Patients who are ultrarapid metabolizers of CYP2D6 may produce more hydromorphone and are more susceptible to hydromorphone toxicity [34].

HYDROMORPHONE

Hydromorphone, a semisynthetic μ-opioid receptor agonist, is structurally similar to morphine. Hydromorphone can be administered orally (both IR and ER form), intravenously, or subcutaneously. IR hydromorphone has a short half-life of 2—3 h requiring administration every 4—6 h for pain management. Moreover, maximum serum levels are achieved within 0.5—1 h. Therefore, ER hydromorphone formulation has been developed where after oral administration, maximum hydromorphone concentration in serum is observed in 3.3—5 h. The mean values of half-life for 2 and 6 mg of the ER tablets under fasting conditions are 8.9 and 16.8 h, respectively. As a result, ER hydromorphone can be administered once daily. The major urinary metabolite is hydromorphone-3-glucuronide [35]. Hydromorphone may have advantage over morphine for pain control. It has much higher bioavailability of 50% compared to morphine. The volume of distribution is 1.22 L/kg [36]. Chemical structure of hydromorphone is shown in Fig. 2.1.

OXYMORPHONE

Oxymorphone is a semisynthetic μ-opioid receptor agonist but may have some δ-opioid receptor agonist effect. Oxymorphone is highly lipophilic, and when administered parenterally, it is approximately 10 times more potent than morphine, but due to poor bioavailability of approximately 10%, oxymorphone is only three times more active than oral morphine. For oral administration, oxymorphone is available both in IR and ER formulations. The half-life of IR formulation is approximately 8 h, while half-life of ER oxymorphone is approximately 10 h. The maximum serum level is observed 2−3 h after administration of ER oxymorphone formulation [37]. Chemical structure of oxymorphone is shown in Fig. 2.1.

Smith et al. reviewed clinical pharmacology of oxymorphone and concluded that oral oxymorphone is more potent (per mg) than oral oxycodone or oral morphine. The maximum serum concentration of oxymorphone is observed within 30 min after administration of IR formulation, but 2−3 h after oral intake of ER form. The plasma protein binding is 10%−12%. The elimination half-life is 7−10 h with IR formulation but 9−13 h with ER formulation. However, steady state is reached after 3 days regardless of the formulation. Oxymorphone undergoes extensive hepatic metabolism via conjugation with glucuronic acid to produce its primary metabolite, oxymorphone-3-glucuronide. Uridine diphosphate glucuronyl transferase subtype B27 catalyzes glucuronidation at the 3 position. A minor metabolite of oxymorphone, 6-hydroxy-oxymorphone, is produced by reduction of 6-keto group. Only less than 2% of the parent compound is recovered in the urine in unchanged form [38].

FENTANYL

Fentanyl is a synthetic full agonist at the μ-opioid receptor and approximately 50−100 times more potent than morphine. Fentanyl was first synthesized by Paul Janssen in 1960 and marketed as a medicinal product for treating pain. Subsequently, many fentanyl analogues were developed, including sufentanil, alfentanil, remifentanil, and carfentanil [39]. Fentanyl is one of the most potent medications known to exist. Similar to other opioid narcotic analgesics, fentanyl's effects include not only analgesia, anxiolysis, euphoria, drowsiness, feelings of relaxation, respiratory depression, constipation, miosis, nausea, pruritus, and cough suppression but also orthostatic hypotension, urinary urgency or retention, postural syncope, and chest wall rigidity especially with IV use. Fentanyl is highly lipophilic and can rapidly cross blood-brain barrier as well as distributes in lipid-rich tissues [40]. Chemical structure of fentanyl is given in Fig. 2.2.

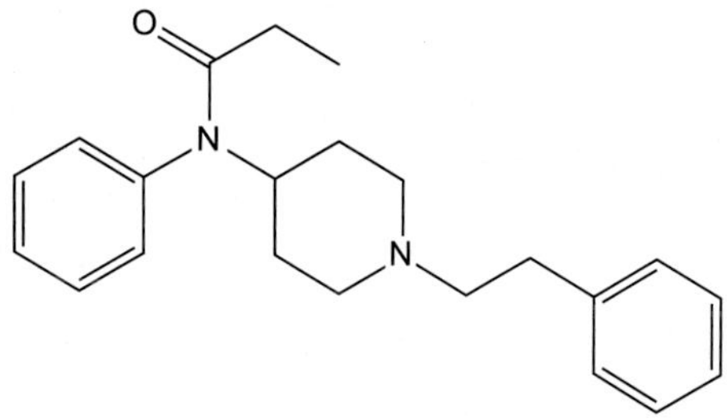

Fentanyl is available as intravenous formulation as well as oral transmucosal formulation like buccal lozenges and transdermal patch, as well as for nasal administration. Fentanyl was approved in 1972 for intravenous administration by the FDA, and in 1990s, fentanyl patches were introduced. Reports of misuse and illicit use by clinicians, primarily anesthesiologists and surgeons with access to the drug, were first reported in the 1980s and continued since then. In 1994, the FDA issued a warning regarding the

FIG. 2.2 Chemical structure of fentanyl.

dangers associated with fentanyl patches, expressing that it should only be prescribed to those with severe pain [41].

Fentanyl is available as citrate salt in an injectable solution containing 50 μg/mL. A single intravenous dose is 25−100 μg fentanyl. Oral transmucosal dosage forms containing 100−1600 μg fentanyl are placed in the mouth for about 15 min at a rate of four doses or less per day. Transdermal fentanyl patches, such as Duragesic patch, contain 1.2−10 mg fentanyl which provides a dose of 12.5−100 μg per hour for 72 h. The onset of action is immediate (within 1.5 min, while peak effect is observed after 4.5 min) after intravenous administration, and duration of action is 30−60 min. After intramuscular administration, the onset of analgesic action is from 7 to 8 minutes. After intravenous administration, terminal half-life of fentanyl is 7 h (range 3−12 h), but terminal elimination half-life may increase up to 16 h in neonates or elderly. The oral formulation is deigned to dissolve slowly in the mouth to facilitate transmucosal absorption. Absolute oral bioavailability of fentanyl is approximately 50%. The analgesic effect may last 2−3 h. Fentanyl is 80%−86% bound to serum proteins and has a volume of distribution of 3−8 L/kg. Fentanyl is metabolized primarily by CYP3A4 isoenzyme. The major metabolite is norfentanyl, which is inactive [41,42]. Minor inactive metabolites of fentanyl include hydroxyfentanyl, hydroxynorfentanyl, despropionyl-fentanyl. These metabolites account for less than 1% fentanyl metabolism in humans. Because fentanyl is metabolized by CYP3A4, inhibitors of CYP3A4 may significantly increase serum fentanyl concentrations causing toxicity. Fetal drug interactions between fentanyl and CYP3A4 inhibitors, fluconazole, clarithromycin, diltiazem, and itraconazole have been reported in chronic pain patients treated with transdermal fentanyl patches. However, other potent CYP3A4 inhibitors such as ritonavir, troleandomycin, itraconazole, voriconazole, and fluconazole showed only a moderate increase in systemic fentanyl exposure. Interaction between fentanyl and ketoconazole has also been reported, where systemic clearance of fentanyl was reduced to 78%. Clinically fentanyl dosage adjustments may become necessary when ketoconazole or other strong CYP3A inhibitors are given simultaneously. Fentanyl itself does not influence CYP3A activity [43].

The effective minimum fentanyl concentration for pain control varies widely between patients (ranging from 0.23 to 0.99 ng/mL). The relationship between plasma fentanyl concentration and pain score is steep, such that small changes in concentration may results in significantly better pain control [44].

In general, fentanyl concentration is not monitored in blood, but the normal level of blood fentanyl after 100 μg/h patch (highest dose available) ranges from 1.9 to 3.8 ng/mL. However, respiratory depression may be detected at levels as low as 1−5 ng/mL. Deaths in patients using fentanyl patches have been reported, where blood concentration varies widely from 3.1 to 43 ng/mL, indicating overlap between therapeutic and toxic concentrations. Biedrzycki et al. reported the accidental overdose of a young black male with sickle cell/β-thalassemia who had been using the Duragesic system for almost 2 years. Toxicological examination revealed blood and urine fentanyl levels of 40 ng/mL and 400 ng/mL, which was 10-fold and 100-fold higher than therapeutic levels [45].

Fentanyl Derivatives

There are several fentanyl derivatives that are used clinically. There are also fentanyl derivatives that are abused but have no clinical use. In this chapter, fentanyl derivatives used clinically are described.

Fentanyl derivatives used in medical or veterinary medicine include alfentanil, carfentanil, remifentanil, and sufentanil. FDA classified these drugs as Schedule II drugs. Alfentanil, a basic drug like fentanyl, is a short active narcotic analgesic that is also an agonist of μ-opioid receptor. Alfentanil is most commonly for short surgical procedures. Alfentanil is usually administered through intravenous infusion, and pharmacological effects can be observed in minutes. The average volume of distribution is 0.39 L/kg and clearance is 0.20 L/kg/h. The elimination half-life is 1.63 h. Children have shorter elimination half-life, smaller volume of distribution, and higher clearance of alfentanil compared to adults. Therefore, children may therefore need a higher rate of infusion for their body size than adults. In the elderly (over 65 years), decreased clearance and prolong elimination half-life have been reported. Plasma protein binding (albumin, lipoproteins, and α-1-acid glycoprotein) is approximately 85% [46].

Alfentanil undergoes extensive CYP3A4 metabolism via two major pathways, forming noralfentanil and N-phenylpropionamide [47].

Remifentanil (remifentanil hydrochloride) is a synthetic opioid and an agonist of μ-opioid receptor, which has a rapid onset of action of about 1 min after initiation of infusion. The drug has a very short half-life, and as a result, steady state is achieved quickly. Although chemically related to the fentanyl family of short-acting 4-anilidopiperidine derivatives commonly used as supplements to general anesthesia, remifentanil is structurally unique among currently available opioids because of its ester linkages. However, remifentanil is not a substrate for plasma cholinesterase and therefore its metabolism is not subject to genetic variance. Remifentanil exhibits a predictable, rapid metabolism by nonspecific esterases in the blood and tissues, principally to a carboxylic acid derivative, remifentanil acid. This organ-independent elimination of remifentanil makes it a useful agent in the intensive care unit setting, where patients commonly have some degree of organ dysfunction. Although metabolism of remifentanil is independent of liver and renal function, its elimination is prolonged in patients with severe renal insufficiency (predicted creatinine clearance <10 mL/min). The volume of distribution of remifentanil in healthy subjects is 0.74 L/kg and clearance is 44.3 mL/min/kg [48]. Remifentanil has a terminal half-life of approximately 10−20 min and its context-sensitive half-time is 3−4 min, regardless of the duration of infusion [49].

Studies indicate that remifentanil may be the first ultrashort-acting opioid that may be used as a supplement to general anesthesia. Compared to alfentanil, the high clearance of remifentanil, combined with its small steady-state distribution volume, results in a rapid decline in blood concentration after termination of an infusion. With the exception of remifentanil's nearly 20-times greater potency remifentanil and alfentanil are pharmacodynamically similar [50].

Sufentanil is even more potent than fentanyl. Alfentanil has the fastest onset of action, followed by sufentanil and then fentanyl. Alfentanil also has the shortest duration of action of the group [51]. After bolus administration, analgesic effect of sufentanil can be observed after 1 min and peak effect is observed after 2.5 min. The volume of distribution is 2.9 L/kg. Elimination half-life of sufentanil is 164 min and total body clearance is 0.762 L/h/kg. Serum protein binding of sufentanil is approximately 92.5% [42].

Carfentanil is a synthetic fentanyl analogue approved for veterinary use. It is a μ-opioid receptor agonist with an estimated analgesic potency approximately 10,000 times that of morphine and 20—30 times that of fentanyl, based on animal studies. Little is known about the pharmacology of carfentanil in humans. However, this drug is abused and its high potency, high lipophilicity, large volume of distribution, and potential active metabolites are concerns for treating people who abuse carfentanil [52].

METHADONE

Methadone is a synthetic opioid that is an agonist of μ-opioid receptor. First synthesized in 1930s, it was approved by the FDA as an analgesic in 1947. Subsequent studies showed that it is effective in treating opioid addiction, and in 1972, FDA approved methadone for treating opioid addiction. Methadone has the longest half-life among all opioids and it is also used as an analgesic in both children and adults. Methadone can be administered orally, intravenously, or other parenteral routes. Methadone is on the World Health Organization list of essential medicines [53]. Chemical structure of methadone is given in Fig. 2.3.

Methadone is utilized for the treatment of moderate to severe pain and detoxification of opiate addiction and methadone maintenance treatment for individuals with opiate dependence. It is administered as a racemic mixture of R- and S-enantiomers, with the R-enantiomer exerting the majority of the pharmacological effect. Methadone is well absorbed following oral dosing with an average bioavailability of approximately 75%. Protein binding of methadone is 71%—88%, and it is mostly bound to α-1-acid glycoprotein. A small fraction of methadone is bound to serum albumin. In a study with 20 opiate addicts receiving 10—225 mg daily, the steady-state plasma levels of methadone ranged from 65 to 630 ng/mL, with peak concentrations of 124—1255 ng/mL [54]. Although there is no difference between R- and S-methadone bioavailability (from 65.5% to 100% for R-methadone and from 65.5% to 100% for S-methadone), R-methadone had a significantly longer mean elimination half-life (37.5 h) than S-methadone (28.6 h) according to one study. In addition R-isomer showed a larger volume of distribution (7.1 L/kg) compared to S-isomer (4.1 L/kg). Clearance was 0.158 and 0.129 L/min for R- and S-methadone, respectively [55].

Methadone is primarily eliminated via hepatic metabolism by cytochrome P450 (CYP) enzymes through oxidative biotransformation. Methadone undergoes stereoselective N-demethylation followed by spontaneous cyclization to form the principle metabolite 2-ethyl-1,5-dimethyl-3,3-diphenylpyrrolidine (EDDP) and 2-ethyl-5-methyl-3,3-diphenylpyrrolidine (EMDP). Both metabolites are inactive. Methadol and normethadol are active analgesic metabolites of methadone, although this pathway is a relatively minor

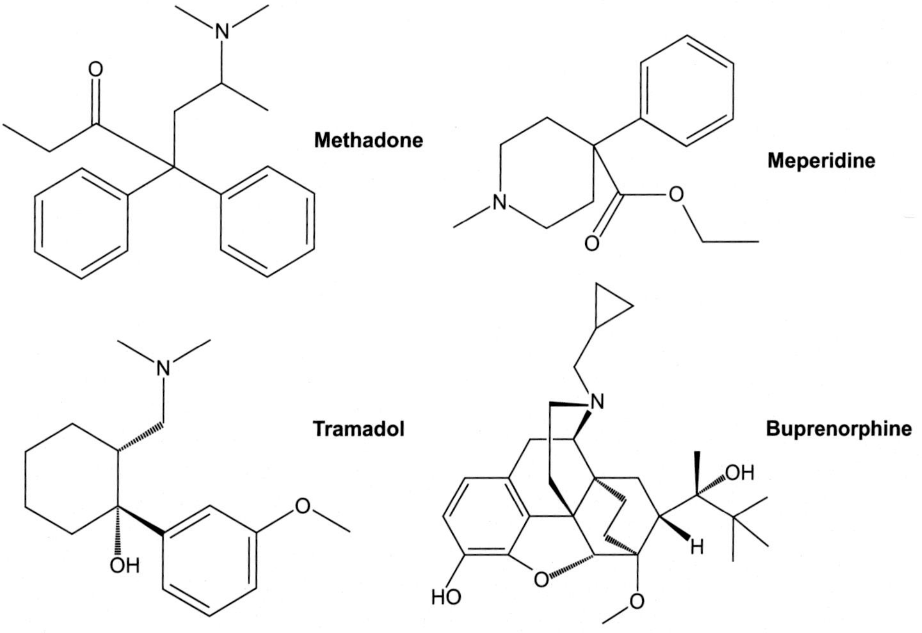

FIG. 2.3 Chemical structures of methadone, meperidine, tramadol, and buprenorphine.

pathway. In addition, to a lesser extent, methadone, normethadol, EDDP, and EMDP are further hydroxylated to form inactive p-hydroxy metabolites. Another minor pathway yields the inactive metabolites 4-dimethylamino-2, 2-diphenylvaleric acid, 4-methylamino-2,2-diphenylvaleric acid, and 1,5-dimethyl-3,3-diphenyl-2-pyrrolidone. The CYP enzymes involved in the formation of methadone metabolites in humans include CYP2B6, 3A4, 2C19, 2D6, and to a lesser extent, CYP2C18, 3A7, 2C8, 2C9, 3A5, and 1A2. The isoenzymes CYP2C19, 3A7, and 2C8 preferentially metabolize (R)-methadone, while CYP2B6, 2D6, and 2C18 primarily metabolize (S)-methadone, but CYP3A4 demonstrates no stereoselectivity in methadone metabolism. Single-nucleotide polymorphisms located within CYPs have the potential to play an important role in altering methadone metabolism and pharmacodynamics of methadone. In addition, methadone's long half-life can be partially attributed to the extreme interindividual pharmacokinetics and its stereoselective metabolism [56].

There is a large interindividual variability in clinical outcomes among methadone maintenance program in patients due to genetic variability. Between 30% and 80% of patients receiving methadone are poor responders. Several factors such as poor coping self-efficacy, mood states, genetic polymorphisms in drug-metabolizing enzymes and transporters, susceptibility to drug interactions, and methadone pharmacokinetics (e.g., clearance, autoinduction) have been suggested as contributing to this variable response in the patients undergoing methadone maintenance program. Methadone patients with withdrawal symptoms have increased systemic exposure of S-methadone versus R-methadone, supporting the concept that S-methadone may be responsible for the negative effects associated with methadone treatment [54].

MEPERIDINE

Meperidine was first synthesized in 1939 as an anticholinergic agent, but it was soon discovered to have analgesic properties [57]. Meperidine (pethidine) is a phenyl piperidine that is also a strong agonist of μ-opioid receptor. Meperidine is distributed in the United States as Demerol. Absorption of meperidine after intramuscular injection is variable and often incomplete. After oral administration, meperidine undergoes hepatic first-pass metabolism, and as a result, only 50%–60% of dose reaches the systematic circulation. However, in patients with hepatic disease, oral bioavailability of meperidine may reach 80%–90%. The volume of

distribution is approximately 3.5 L/kg. Meperidine is approximately 65%–75% bound to serum proteins. Meperidine is metabolized by the liver, where hydrolysis of meperidine produces meperidinic acid and N-demethylation results in the formation of normeperidine. Normeperidine also undergoes hydrolysis producing normeperidinic acid. The half-life of meperidine is 2.5–4 h after therapeutic dosage. Normeperidine is a CNS stimulant, and CNS toxicity following therapy with meperidine may be related to normeperidine [58]. Todd et al. based on a study of 11 term pregnant women reported that average termination half-life of methadone was 13.3 h, which is much longer than half-life reported by other authors. The authors also reported that after intravenous administration of meperidine, peak serum concentration of meperidine was observed after 5 min with a rapid decrease of one-third after 15 min. However, measurable meperidine level in serum was observed up to 48 h. Mean plasma clearance in pregnant women was 0.678 L/min [59]. Chemical structure of meperidine is given in Fig. 2.3.

Although meperidine is FDA approved for relieving moderate to severe pain and has been widely used since its introduction in the 1930s, currently some clinicians do not consider this drug as a first-line opioid analgesic due to concerns about adverse reactions, drug interaction, and normeperidine neurotoxicity. Nevertheless, meperidine should be considered appropriate choice for acute relief of moderate to severe pain in patients who cannot tolerate other opioids or also in patients who do not respond properly to other opioids. Moreover, oral administration of meperidine should be discouraged because of extensive first-pass metabolism by the liver. In addition, oral administration of meperidine results in higher normeperidine concentrations when compared with parenteral administration. Therefore, risk of normeperidine-induced neurotoxicity is higher with oral administration of meperidine [60].

TRAMADOL

Tramadol is an opioid drug that, unlike classic opioids, also modulates the monoaminergic system by inhibiting noradrenergic and serotoninergic reuptake. For this reason, tramadol is considered an atypical opioid. These special pharmacological characteristics have made tramadol one of the most commonly prescribed analgesic drugs to treat moderate to severe pain [61]. Tramadol was first developed in Germany in 1970s and was approved by the FDA in 1995 but was classified as Schedule IV drug in 2014 [62]. Chemical structure of tramadol is given in Fig. 2.3.

Tramadol, a centrally acting analgesic structurally related to codeine and morphine, consists of two enantiomers, both of which contribute to analgesic activity via different mechanisms. Tramadol is available as a racemic mixture of both R (+) and S (−) tramadol. The (+) tramadol and the metabolite (+)-O-desmethyl tramadol (M1) are agonists of the μ-opioid receptor. In addition, (+) tramadol is the more potent serotonin reuptake inhibitor, while the (−) tramadol inhibits norepinephrine reuptake, enhancing inhibitory effects on pain transmission in the spinal cord. The complementary and synergistic actions of the two enantiomers improve the analgesic efficacy and tolerability profile of the racemate. Tramadol is available in various pharmaceutical preparations, including ampules containing 50 mg (1 mL) or 100 mg (2 mL) of tramadol in a solution for intravenous, intramuscular, or subcutaneous injection. IR oral capsules usually contain 50 mg tramadol and should be taken four to six times daily for adequate pain control. Sustained-release formulations (100, 150, or 200 mg tramadol) may be taken once or twice daily. After oral administration, tramadol is rapidly and almost completely absorbed. Peak concentration is observed 1.6−1.9 h after oral administration of tramadol capsule. After administration of 100 mg tramadol orally, peak plasma level is approximately 300 ng/mL. Tramadol is also rapidly absorbed after intramuscular administration, and peak serum concentration of 166 ng/mL has been reported at 0.75 h after intramuscular injection of 50 mg tramadol. Sustained-release tablets release the active ingredient over a period of 12 h, reaching peak concentrations after 4.9 h and have a bioavailability of 87%−95% compared to capsules. Tramadol is rapidly distributed in the body, and plasma protein binding is approximately 20% [63].

Tramadol is mainly metabolized by O- and N-demethylation and by conjugation reactions forming glucuronides and sulfates. Tramadol and its metabolites are mainly excreted via the kidneys. The mean elimination half-life is about 6 h. The O-demethylation of tramadol to M1, the main analgesic effective metabolite, is catalyzed by CYP2D6 isoenzyme, whereas N-demethylation to M2 is catalyzed by CYP2B6 and CYP3A4. The wide variability in the pharmacokinetic properties of tramadol can partly be ascribed to CYP polymorphism. O- and N-demethylation of tramadol and renal elimination are stereoselective. The analgesic potency of tramadol is about 10% of that of morphine following parenteral administration. Tramadol provides postoperative pain relief comparable with that of meperidine, and the analgesic efficacy of tramadol

can further be improved by combination with a nonopioid analgesic. Tramadol may prove particularly useful in patients with a risk of poor cardiopulmonary function, after surgery of the thorax or upper abdomen, and when nonopioid analgesics are contraindicated. Tramadol is an effective and well-tolerated agent to reduce pain resulting from trauma, renal or biliary colic, and labor and also for the management of chronic pain of malignant or nonmalignant origin, particularly neuropathic pain. Tramadol appears to produce less constipation and dependence than equianalgesic doses of strong opioids [63].

Metabolism of tramadol is very complex. Wu et al. studied metabolism of tramadol hydrochloride in three male volunteers after single oral administration of 100 mg of the drug. The authors identified unchanged tramadol and a total of 23 metabolites, consisting of 11 Phase I metabolites (M1-11) and 12 conjugates (7 glucuronides, 5 sulfate-conjugated metabolites), in urine using mass spectrometry [64].

TAPENTADOL

Tapentadol is structurally related to tramadol and was approved by the FDA in 2008. Tramadol and tapentadol are two centrally acting synthetic opioids with an atypical mechanism of action by acting as μ-opioid receptor agonist and also as norepinephrine reuptake inhibitors (MOR-NRI agents). Tramadol is a prodrug that acts through noradrenaline and serotonin reuptake inhibition, with a weak opioid component added by its metabolite O-desmethyl tramadol. In contrast, tapentadol does not require metabolic activation and acts mainly through noradrenaline reuptake inhibition and has a strong opioid activity. Such features confer tapentadol potential advantages, namely lower serotonergic, dependence and abuse potential, more linear pharmacokinetics, greater gastrointestinal tolerability, and applicability in the treatment of chronic and neuropathic pain over tramadol. However, in vivo and in vitro studies have shown that tramadol and tapentadol cause similar toxicological damage. In this context, it is important to underline that the choice of opioid should be individually balanced and a tailored decision, based on previous experience and on the patient's profile, type of pain, and context of treatment [65]. Tapentadol is effective in managing both nociceptive and neuropathic pain; it is approved for treating painful diabetic peripheral neuropathy and has been shown to be effective in treating other neuropathic pain and phantom limb pain. Tapentadol is available in IR and prolonged-release formulations [66].

Tapentadol is 32% absorbed after oral administration. Since food has no effect on its oral bioavailability, tapentadol may be taken with or without food. Gastric pH and gastric motility do not affect its absorption. Tapentadol readily crosses the blood-brain barrier and is widely distributed throughout the body. The plasma half-life is approximately 4 h after oral administration. Maximum serum concentration of tapentadol is observed 1 h after oral administration. The duration of analgesic effect is 6 h. Tapentadol undergoes hepatic first-pass metabolism. A small amount is metabolized by Phase I pathways, which forms hydroxy tapentadol (hydroxylation pathway catalyzed by CYP2D6) and N-desmethyl tapentadol (N-demethylation pathway catalyzed by CYP2C9 and CYP2C19). The hydroxy tapentadol represents on 2% of metabolism, while N-desmethyl tapentadol represents 13% of the metabolism. However, major metabolic pathway is Phase II metabolism producing glucuronide conjugate (55%) and sulfate conjugate (15%). Only 3% of drug is excreted as unchanged form of drug in urine. Both tapentadol and its metabolites are excreted mainly (99%) through the kidney. Because Phase I metabolism plays a minor role in metabolism of tapentadol, drug-drug interactions are minimal [67].

BUPRENORPHINE

Buprenorphine is a semisynthetic opiate that is synthesized from natural alkaloid thebaine found in opium. Buprenorphine was introduced in the early 1980s as an opioid analgesic in Europe and subsequently for the treatment of opioid addiction in France in 1996. Buprenorphine was approved by the US FDA in October 2002 as a Schedule III narcotic for use in treating opioid-dependent men and opioid-dependent women who are not pregnant. Buprenorphine has a significantly different mechanism of interaction with opioid receptors compared with other opioids such as morphine. Buprenorphine is a potent but partial agonist of μ-opioid receptor, showing a high affinity but low intrinsic activity. High potency and slow off rate (half-life of association/dissociation is 2−5 h) help buprenorphine displace other μ-opioid agonists such as morphine from receptors and overcome opioid dependence issues. Buprenorphine is approximately 25−100 times more potent than morphine and has a prolonged therapeutic effect that is very useful to opioid dependance, as well as pain. Interestingly, buprenorphine is a potent κ-receptor antagonist, as well as an antagonist for δ-opioid receptors. Buprenorphine is a

preferred opioid for treatment of pain in patients with renal or liver dysfunction [68]. Chemical structure of buprenorphine is given in Fig. 2.3.

Buprenorphine is a lipophilic chiral molecule and has low solubility in water. Compared to 100% bioavailability of buprenorphine after intravenous administration, bioavailability is 49% after administration of sublingual solution and 29% with sublingual tablets. Sublingual and transdermal formulations tend to show long half-life (20−73 h). With a sublingual formulation, buprenorphine shows onset of effects at 30−60 min after dosing, and the peak clinical effects are observed at 1−4 h. The duration of effect may last for 6−12 h at low dose (<4 mg) and 24−72 h at higher dose (>16 mg). The longer effect at higher buprenorphine sublingual dose may be linked to sustained, effective drug levels for extended duration because of its slower elimination and enterohepatic recirculation. Buprenorphine has a large volume of distribution, and the drug is strongly protein bound (approximately 96%) mostly to α- and β-globulins. Serum buprenorphine concentration is usually low. Buprenorphine is extensively metabolized in the liver, and the major metabolite, norbuprenorphine, is formed by CYP3A4-mediated N-dealkylation of buprenorphine. In addition, buprenorphine is conjugated with glucuronic acid during Phase II metabolism by the action of UGT2B7 enzyme, while norbuprenorphine is also conjugated with glucuronic acid by the enzyme UDP-glucuronosyltransferase 1A1 in the liver. Buprenorphine is eliminated primarily via a stool (as free forms of buprenorphine and norbuprenorphine), while 10%−30% of the dose is excreted in urine as conjugated forms of buprenorphine and norbuprenorphine. The plasma levels of conjugate metabolites buprenorphine-3-glucuronide and norbuprenorphine-3-glucuronide can exceed the parent drug levels [68].

LEVORPHANOL

Levorphanol was synthesized in 1949 and then approved by the FDA in 1953 for treating moderate to severe pain. Levorphanol is a unique synthetic opioid due to its wide range of activities, including agonist for μ-opioid receptor, as well as both δ- and κ-opioid receptors. In addition, levorphanol is an antagonist of N-methyl-D-aspartate receptor, as well as reuptake inhibitor of both norepinephrine and serotonin. This multimodal profile might prove effective for pain syndrome that is refractory to other opioid analgesics, such as central and neuropathic pain and opioid-induced hyperalgesia. Levorphanol is well suited as a first-line opioid and can also be used during opioid rotation. It has no known effect on the cardiac QT interval or drug-drug interactions involving hepatic CYPs enzymes. In these regards, levorphanol may offer a superior safety profile over methadone and other long-acting opioids. However, this drug is less frequently prescribed than other opioids and is considered as a forgotten opioid [69].

Levorphanol can be administered orally, intermuscularly, or as intravenous bolus. Usual oral dose is 2 mg levorphanol tartrate. After oral administration peak concentration is observed in 1 h, and the terminal half-life is 11–16 h. The analgesic effects lasts for 6–15 h, thus levorphanol is considered as a long-acting opioid. Analgesic effect is observed with a serum levorphanol concentration of 10 ng/mL. The serum protein binding is approximately 40%. Levorphanol is metabolized to an inactive metabolite levorphanol-3-glucuronide. The glucuronide metabolite concentration in serum may reach 5–10 times greater than concentration of the parent drug. Glucuronide metabolite is eliminated renally [69].

PROPOXYPHENE

Propoxyphene is a short-acting, Schedule IV, synthetic opioid, which was introduced in 1957 for treating mild-to-moderate pain. Propoxyphene was available as the single drug (Darvon 65 mg) or in combination with other analgesics, most commonly with acetaminophen. When compared to other analgesics in acute and chronic pain, propoxyphene with acetaminophen was consistently shown to be equivalent or inferior to other commonly used analgesics. In postoperative pain alone, propoxyphene with acetaminophen was found less effective than 400 mg ibuprofen for pain relief. However, propoxyphene also showed serious safety concerns. The mostly widely recognized serious side effect is cardiotoxicity leading to potentially fatal arrhythmias. Moreover, propoxyphene was linked to more drug-related deaths than either hydrocodone or tramadol and more emergency department visits than codeine. Despite the potential risks of propoxyphene and nonsuperior analgesic properties relative to common alternatives, propoxyphene with acetaminophen was a commonly used drug in the United States for a long period of time. However, due to accumulating data regarding cardiotoxicity and following previous action by European regulatory agencies to withdraw propoxyphene from their markets, FDA requested that propoxyphene be removed from the US market, and on November 19, 2010, Xanodyne Pharmaceuticals agreed to remove propoxyphene from the US market [70].

EFFECT OF SEX, AGE, ETHNICITY, AND REANAL AND HEPATIC FUNCTION ON CERTIN OPIOIDS

Only oxycodone and hydromorphone show sex difference in serum drug level, where concentrations are approximately 25% higher in women than men for oxycodone and maximum serum concentrations of hydromorphone are approximately 25% higher in women. However, dosage adjustments are needed for morphine, codeine, hydrocodone, methadone, and fentanyl in elderly patients mostly due to reduced clearance. Steady-state concentration of oxymorphone is approximately 40% higher in patients older than 65 years. Dosage adjustments also needed for morphine, codeine, oxycodone, methadone, tramadol, and hydromorphone. Hydrocodone is also commonly administered with acetaminophen, therefore liver function must be considered before initiating therapy with hydrocodone acetaminophen combination. Oxymorphone is contraindicated in patients with moderate to severe liver impairment. Although there is no effect on ethnicity on pharmacokinetic parameters of oxycodone, methadone, tramadol, fentanyl, hydromorphone, and oxymorphone, Chinese patients have higher clearance for morphine. For codeine and hydrocodone, CYP2D6 allelic variations observed in Asian as well as African dissents may have significant effect on metabolism [71].

OTHER OPIOIDS

In addition to commonly prescribed opioids described above, other opioids are also used in pain management. Ethylmorphine is a cough suppressant and an opioid analgesic. Ethylmorphine is metabolized by N-demethylation to norethylmorphine (catalyzed by CYP3A4) and by O-deethylation to morphine (catalyzed by CYP2D6) [72]. Dihydromorphine differs structurally from morphine only by the reduction of the 7,8-double bond. Binding assays have confirmed its high selectivity for μ-opioid receptors, and it has long been considered pharmacologically equivalent to morphine, although slightly more potent [73]. Dihydrocodeine is a semisynthetic opioid that is frequently used as an analgesic and antitussive drug. Furthermore, doses up to 2500 mg/day are prescribed in substitution therapy for heroin-addicted people. Dihydrocodeine is structurally related to the naturally occurring opioid codeine. It undergoes O-demethylation to dihydromorphine (catalyzed by CYP2D6) [74]. Interestingly, analgesic effect following dihydrocodeine ingestion is mainly attributed to the parent drug rather than its dihydromorphine metabolite. It can thus be inferred that polymorphic differences in dihydrocodeine metabolism to dihydromorphine have little or no effect on the analgesic affect [75].

Pentazocine not only has analgesic effect compared to morphine but also possesses weak opiate antagonistic activity that is likely to render it less liable to dependance than morphine. Following intravenous administration of 35 mg pentazocine, the man plasma half-life was 135 min with a peak blood level of 200 ng/mL. The half-life was 120 min after 45 mg intramuscular injection and peak plasma level of 100–243 ng/mL was observed 15–60 min after injection. After oral administration of 75 mg pentazocine, peak concentrations of 110–300 ng/mL were observed between 1 and 3h. However, blood levels of pentazocine were less predictable after oral administration due to significant variations between different patients. Pentazocine is extensively metabolized to inactive metabolites that are excreted in urine. Therefore, pentazocine should be administered with caution in patients with renal failure [76].

Butorphanol is a synthetic opioid that is pharmacologically related to morphine but more potent than morphine. It is used to manage different kinds of pain (postoperative, labor and delivery, migraine, etc.), as well as a perioperative medicine and supplement to anesthesia. Initially, this drug was only available for parenteral administration due to poor bioavailability (5%–17%) after oral administration. Later, an intranasal formulation was developed (butorphanol tartrate; 1 or 2 mg single dose) to bypass the first-pass effect, thus allowing a higher systemic absorption. Mean systemic availability of intranasal butorphanol is 50%–70%. Median time to reach maximum serum level after intranasal administration of butorphanol is 15–30 min [77].

Nalbuphine is an opioid agonist-antagonist of the phenanthrene series, which was synthesized in an attempt to provide analgesia without the undesirable side effects of the pure agonists such as morphine. This drug is indicated for treating mild-to-moderate pain and is also effective for the treatment of conditions ranging from burns, multiple trauma, orthopedic injuries, gynecology, and intraabdominal conditions. Its analgesic and possibly certain antipruritic effects are mediated via actions on the μ- and κ-opioid receptors [78]. Nalbuphine pharmacokinetics is moderately altered in renal failure. Although available for almost 40 years, nalbuphine has drawbacks: it is not an oral formulation, it causes withdrawal in patients on sustained released opioids, and it cannot be used to treat an opioid withdrawal syndrome. Nalbuphine, despite

being a μ-opioid receptor antagonist, produces a drug-liking effect and can be abused. However, there are very few deaths associated with nalbuphine alone in part due to the fact that it is not only rarely used but also related to a ceiling on respiratory depression [79].

Pharmacological strategies for curbing symptoms of opioid withdrawal is based on opioid replacement therapy in which the primary goal is to substitute short-acting opioid with long-acting and less euphoric opioids. The driving principle is that opioids with a long half-life produce stable drug plasma levels, circumventing the need for frequent dosing. Methadone is widely used in treating patients with opioid dependance. However, methadone maintenance program is limited by a shortage in the number of doctors licensed to prescribe methadone. Also, most programs require patients to pick up methadone prescriptions on a day-to-day basis to limit the risk of abuse or overdose, which can introduce significant barriers to individuals living in remote locations or with jobs that require frequent travel. Buprenorphine/naloxone (Suboxone) has emerged as an alternative option to methadone for opioid replacement therapy. The combination of buprenorphine, a partial μ-opioid receptor agonist, and naloxone, an opioid receptor antagonist, has conferred several therapeutic advantages over methadone for curbing opioid withdrawal, including an improved safety profile and reduced risk of overdose-related death [80].

Naloxone is probably the most commonly used medication for treating a patient overdosed with an opioid. Naloxone, an opioid antagonist, was first approved by the FDA in 1971 for intravenous administration in patients for reversal of opioid overdose. Naloxone has a high affinity of μ-opioid receptor and can rapidly reverse opioid overdose. However, in patients with no overdose of opioids, binding of naloxone with opioid receptors has no adverse clinical effect. Naloxone distribution program exists in many states where nonmedical persons may administer naloxone. New formulations of naloxone are continuously introduced, and in August 2014, the FDA approved a new formulation of naloxone for use as an intramuscular autoinjector known as Evzio (4 mg naloxone). A new intranasal formulation was approved by the FDA in November 2015. The nasal spray dose is 4 mg, which should be administered into one nostril. Regardless of route of administration, naloxone is very effective in reversing opioid overdose [81].

Vanky et al. studied pharmacokinetics of naloxone (10 μg/kg) after intranasal administration using 20 healthy volunteers. Median time to maximum naloxone concentration was 14.5 min, and mean maximum naloxone concentration (Cmax) was 1.09 ± 0.56 ng/mL. Elimination half-life estimated from the median concentration data was 28.2 min [82]. In another study, the authors compared pharmacokinetic parameters after intravenous, intramuscular, and intranasal administration of naloxone in healthy subjects. Plasma concentration following intravenous administration (0.4 mg) reached an early peak (geometric mean Cmax 5.94 ng/mL, median and maximum serum levels were observed in approximately minutes), followed by a rapid decline during the next 10 min, and a gradual decline thereafter. The intramuscular administration (0.4 mg) showed more gradual early uptake, with lower and later peak (geometric mean Cmax 1.27 ng/mL, median time for maximum serum levels: 10 min), and flatter and slower decline thereafter. However, after intranasal administration, dosage (1 mg, 2 mg or 4 mg) serum levels were higher than 0.4 mg intravenous or intramuscular administration [83].

Naltrexone blocks dopamine that makes this drug a potential alternative to opioid replacement therapy for opioid-dependent patients. Naltrexone is used as a relapse prevention approach in patients already enrolled in abstinence-oriented programs because patients addicted to opioids can no longer get high while taking naltrexone. In addition, risk of overdose is very low. However, naltrexone has limited use in treating opioid-dependent patients because it cannot be given to a patient who has been using opioid without a 7–10 days opioid-free period (washout phase) because it may precipitate and have acute withdrawal symptom due to its antagonist effect [84].

ABUSE-DETERRENT OPIOID FORMULATIONS

Opioid abuse can be classified under three categories:
- Intentional abuse
- Therapeutic errors
- Accidental overdose

Commonly, the majority of chronic intentional opioid abusers tamper with tablets, so that most of the drug is available for achieving a high opioid concentration in blood and hence euphoria. ER formulations are favored by abusers as they contain a high dose of active opioid, and ER properties are readily neutralized by crushing or dissolving. An intervention that addresses the public health implications of nonmedical opioid abuse is the development of an abuse-deterrent formulation (ADF) without compromising characteristics of pharmaceutical products e.g., chemical stability, correct release rate, reproducible manufacture at commercial scale, and clinical efficacy and safety [85]. In general, hydrocodone, oxycodone, and oxymorphone are often abused by inhalation, whereas hydromorphone and morphine are more prone to be abused by injection [86].

There are different types of methodologies to formulate abuse-deterrent opioid drugs:
- Physical/chemical barrier technology.
- Combining an opioid drug with an opioid antagonist.
- Incorporation of an aversive ingredient intended to induce unpleasant symptoms following abuse by the nasal route.
- Prodrug that must be taken orally because enzymatic cleavage of prodrug to release the active ingredient in the gastrointestinal tract is needed for its pharmacological action. The prodrug is inactive if snorted or administered intravenously.
- Alternative delivery system such as dermal patch.

These methodologies along with example of opioid formulation following such approaches are summarized in Table 2.8.

Physical/Chemical Barrier Technology

Oxycodone is highly abuse, and in 2010, FDA approved ADF of oxycodone as OxyContin (ERO), which was oxycodone reformulated with a polymer matrix providing a physical barrier to tablet breaking and crushing. Moreover, when exposed to water, this reformulation turns into a viscous gel that resists being drawn into a syringe and subsequent injection. Studies have shown bioequivalence of this product with original oxycodone formulation. Moreover, this reformulation also reduced oxycodone abuse. Opana (oxymorphone) has been reformulated to incorporate a polyethylene oxide matrix designed to render the tablet highly resistant to crushing without affecting its ER properties. Moreover, when exposed to water, this reformulation forms a gel, which is difficult to draw into a syringe. This formulation has shown to be bioequivalent to the original ER oxymorphone formulation leading to FDA approval of its distribution in December 2011 [85]. In postmarketing studies, it was observed that the reformulation was more difficult to crush into particles suitable for insufflation or injection, resulting in the great majority of study participants reporting unwillingness to insufflate or inject the tampered product [87].

A new formulation of abuse-resistant ER morphine tablet (Morphine ARER: MorphaBond) contains inactive ingredients that inhibit physical manipulation or chemical extraction without interfering with ER morphine

TABLE 2.8
Methodologies to Formulate Abuse-Deterrent Opioids.

Methodology	Comments	Example
Physical/chemical barrier	There is no risk if used as prescribed and bioequivalence between extended release and abuse-deterrent extend release formulation has been established.	• OxyContin (ERO) • Hyder (oxycodone bitartrate) • Xtampza ER (oxycodone) • Opana (oxymorphone) • Nucynta (tapentadol) • MorphaBond (morphine sulfate pentahydrate) • Arymo ER (morphine sulfate) • Vantrela (hydrocodone bitartrate) • Hysingla (hydrocodone bitartrate)
Opioid drug formulated with an antagonist	No effect if formulation is manipulated for abuse.	• Suboxone (buprenorphine/naloxone combination) • Embeda (morphine sulfate with naltrexone) • Targiniq ER (oxycodone with naloxone) • Toxyca ER (oxycodone with naltrexone)
Aversion	Prevent abuse by crushing, chewing, or using nonoral route of abuse.	• Oxecta (oxycodone)
Prodrug formulation	Medication is activated by enzymes in the gastrointestinal tract and has no effect if abused using nonoral route.	• Apadaz (acetaminophen and benzhydrocodone hydrochloride; previously KP201/APAP), approved by the FDA in 2018
Novel delivery system	Avoid opioid abuse.	• Buprenorphine transdermal delivery • Buprenorphine subdermal implant (probuphine)

characteristics. A viscous material is produced if tablet is attempted to dissolve in liquid, thus preventing abuse by injection. The formula is also resistant to physical manipulation such as cutting, crushing, or breaking [88]. FDA approved this formulation in 2015.

In 2011, the FDA approved ER tapentadol (Nucynta) formulated with a polyethylene oxide matrix (physical barrier) that confers resistance to crushing or extraction of the active drug [85]. Vosburg et al. based on their study concluded that tamper-resistant formulation of tapentadol tablets does not appear to be well-liked by individuals who tamper regularly with extended-release oxycodone tablets. Employing tamper-resistant technology may be a promising approach toward reducing the abuse potential of tapentadol extended-release [89].

A new oxycodone abuse-deterrent formula (Xtampza) was approved by the FDA in 2016, which utilizes DETERx microsphere in capsule technology. When such formulation is dissolved, microspheres combine with fatty acids and waxes in the capsule, thus preventing abuse by chewing, crushing, or extraction for intravenous administration. However, this formulation must be

taken with food because its absorption is increased by high-fat meal. In January 2017, FDA approved two more ADF: Arymo ER (morphine sulfate ER) and Vantrela ER (hydrocodone bitartrate ER) [86].

Combining an Opioid Drug with an Opioid Antagonist

An example of such approach is Suboxone (buprenorphine/naloxone combination), where buprenorphine is combined with naloxone, an opioid antagonist. This combination was approved by the FDA in 2003 for the treatment of opioid dependence but is commonly used off-label as a drug for pain management. It is available as both a sublingual tablet and film. In two independent studies, this formulation demonstrated reduced abuse liability in comparison to standard buprenorphine tablets.

Embeda (ER morphine/naltrexone combination), where morphine is combined with naltrexone, an opioid antagonist, is another example of opioid agonist/antagonist combination. Although unlike naloxone, naltrexone is absorbed after oral intake, naltrexone is not released when this formulation is taken orally as prescribed because morphine is released over an extended period of time, whereas the naltrexone is sequestered in the inner core. However, the tablet is chewed or crushed and the naltrexone is released with the morphine where opioid-dependent individuals can incur symptoms of withdrawal if this medication is not taken as prescribed [85].

Targiniq ER (oxycodone ER combined with naloxone in a fixed 2:1 ratio) is another example of such formulations. In 2016, FDA approved Troxyca ER (oxycodone ER combined with naltrexone). In this formulation, pellets of oxycodone hydrochloride are surrounded by naltrexone. If the tablet is crushed, effect of oxycodone will be neutralized by naltrexone [86].

Incorporation of Aversive Ingredients

An example of incorporation of aversive ingredient in an opioid formulation in order to deter abuse is Oxecta where IR oxycodone is combined with sodium lauryl sulfate (to irritate nasal passage), as well as components that form a gel when exposed to water in order to deter intravenous administration. The product was approved by the FDA in 2011 [85]. Studies have shown that this formulation is less abused than regular oxycodone [86].

Abuse-Deterrent Prodrug Formulation

Prodrug contains an inactive drug that must be converted by enzymes in the gastrointestinal tract to form a pharmacologically active opioid. Therefore, such formulation cannot be abused by inhalation, intravenous injection, or nonoral route. Benzhydrocodone is a prodrug of hydrocodone where a benzoic acid is attached covalently to hydrocodone molecule. This prodrug must be converted into hydrocodone in the gastrointestinal track for pharmacological activities. A combined benzhydrocodone and acetaminophen formulation (IR) has bene developed as an ADF that received FDA approval in February 2018 (Apadaz) for the short-term (no more than 14 days) management of acute pain severe enough to require an opioid analgesic and where alternative treatments are inadequate. This formulation is bioequivalent to regular hydrocodone/acetaminophen combination [90].

Novel Drug Delivery System

Another approach to ADF is to use subcutaneous implant such as buprenorphine transdermal delivery system. Transdermal buprenorphine is available (5, 10, and 20 µg/h delivery of buprenorphine) as transdermal patches for administration once every 7 days (BuTrans). These are approved in Europe, the United States, Japan, and various other countries for the management of chronic pain. The absorption of buprenorphine from the transdermal patch may be affected by the application site. The recommended sites for application of transdermal buprenorphine are the upper outer arm, upper chest, upper back, and the side of the chest. Buprenorphine diffuses from the patch through the skin into the circulation, providing a constant delivery of buprenorphine over the 7-day dosing interval. The advantage of patch over sublingual buprenorphine is in the known constant delivery of drug and the once a week dosing requirement. After administration of transdermal buprenorphine, the plasma buprenorphine concentration gradually increases over the first 2 days and then reaches a plateau. A 10 µg patch produces an average serum buprenorphine level of 155 pg/mL. This is a significant improvement over sublingual buprenorphine, which shows large fluctuations between peak and trough plasma drug concentrations in any 24-h period. In addition to pain management, transdermal buprenorphine may be effective in management of opioid dependence with superiority over sublingual buprenorphine in pharmacokinetics and tolerability with the added advantage of once a week administration [91].

Implantable buprenorphine (Probuphine) is approved for the treatment of opioid-dependent patients who have achieved and sustained clinical stability on a transmucosal buprenorphine product at low-to-moderate dosages, defined as no more than 8 mg per day. It is intended for the treatment of opioid-dependent patients who have achieved and sustained clinical stability on a transmucosal buprenorphine product at low to moderate dosages, defined as no more than 8 mg per day. Probuphine can be prescribed, inserted, and removed only by healthcare professionals who have received certification and training through the FDA's Probuphine Risk Evaluation and Mitigation Strategy program. This product should not be used in pregnant women because of the inflexibility of dosing. Serious complications are related to improper insertion and removal of the buprenorphine implant, as well as improper initiation in patients taking other types of opioid therapy who are not already stabilized on a buprenorphine regimen. Implantable buprenorphine is an expensive alternative for patients (1 month's supply of four patches cost over $5000) on a stable dosage of transmucosal buprenorphine. Based on current research and guidance, implantable buprenorphine should not be used as initial treatment for opioid dependence or as a substitute for medication-assisted treatments other than transmucosal buprenorphine [92].

COMPLIANCE WITH PRESCRIPTION OPIOID

Because opioid abuse is a serious public health issue, a validated screening approach that provides an effective and rational method of selecting patients for opioid therapy, predicting risk, and identifying problems once they arise would be very useful in clinical practice. Such approach could potentially curb the risk of iatrogenic addiction. Multiple opioid assessment screening tools and approaches have been developed by various authors. In addition, urine drug testing, monitoring of prescribing practices, prescription monitoring programs, opioid treatment agreements, and utilization of universal precautions are essential [93]. Urine drug testing in pain management patients is very useful in identifying patients who may misuse their opioid medications. Please see Chapter 7 for more detail.

CONCLUSIONS

As mentioned earlier, over 100 million American people live with chronic pain. Therefore, proper pain management is essential for well-being of these patients. Although opioid medications are very effective in controlling pain, nonprescription abuse of opioid is a very serious problem not only because it may cause not only opioid dependance but also significant opioid-related overdoses and fatality in some opioid-overdosed patients. In order to combat opioid epidemic, pharmaceutical companies have developed abuse-deterrent formula that are bioequivalent to original opioid formulation but have less desirability on opioid abusers. Keast et al. based on their study concluded that both prescription spending and physician and pharmacy spending combined may be increased with the use of these new products because of higher pricing. However, for patients with comorbidities of addiction, using ADFs may lower the overall medical costs due to lower abuse potential of these formulations [94].

ACKNOWLEDGMENT

All structures presented in this chapter are drawn by Mathew D. Krasowski, MD, PhD, Vice Chair of Clinical Pathology, University of Iowa, Roy J., and Lucille A., Caver College of Medicine, Iowa City, IA. Figures are courtesy of Dr. Krasowski.

REFERENCES

[1] Tompkins DA, Hobelmann JG, Compton P. Providing chronic pain management in the "Fifth Vital Sign" Era: historical and treatment perspectives on a modern-day medical dilemma. Drug Alcohol Depend 2017; 173(Suppl. 1):S11−21.

[2] Johannes CB, Le TK, Zhou X, Johnston JA, Dworkin RH. The prevalence of chronic pain in United States adults: results of an Internet-based survey. J Pain 2010;11: 1230−9.

[3] Shmagel A, Foley R, Ibrahim H. Epidemiology of chronic low back pain in US adults: data from the 2009−2010 National Health and Nutrition Examination survey. Arthritis Care Res 2016;68:1688−94.

[4] McWilliams LA, Cox BJ, Enns MW. Mood and anxiety disorders associated with chronic pain: an examination in a nationally representative sample. Pain 2003;106: 127−33.

[5] Gaskin DJ, Richard P. The economic costs of pain in the United States. J Pain 2012;13:715−24.

[6] Meldrum M. A capsule history of pain management. J Am Med Assoc 2003;290:2470−5.

[7] Stein C. Opioid receptors. Annu Rev Med 2016;67: 433−51.

[8] Bernstein MH, Beaudoin FL, Magill M. Response to FDA Commissioner's statement on Dsuvia approval. Addiction 2019;114:757−8.

[9] Mandell BF. The fifth vital sign: a complex story of politics and patient care. Cleve Clin J Med 2016;83:400−1.

[10] Schug SA, Zech D, Grond S, Jung H, et al. A long-term survey of morphine in cancer pain patients. J Pain Symptom Manag 1992;7:259−66.

[11] Perry S, Heidrich G. Management of pain during debridement: a survey of U.S. burn units. Pain 1982;13:267−80.

[12] Battista NA, Pearcy LB, Strickland WC. Modeling the prescription opioid epidemic. Bull Math Biol 2019;81: 2258−89.

[13] Van Zee A. The promotion and marketing of oxycontin: commercial triumph. Am J Publ Health 2009;99(2): 221−7.

[14] Scholl L, Seth P, Kariisa M, Wilson N1, Baldwin G. Drug and opioid-involved overdose deaths — United States, 2013−2017. MMWR Morb Mortal Wkly Rep 2018; 67(51−52):1419−27.

[15] Guy Jr GP, Zhang K. Opioid prescribing by specialty and volume in the U.S. Am J Prev Med 2018;55:e153−5.

[16] Nataraj N, Zhang K, Guy Jr GP, Losby JL. Identifying opioid prescribing patterns for high-volume prescribers via cluster analysis. Drug Alcohol Depend 2019;197:250−4.

[17] Liu T, Ivaturi V, Gobburu J. Integrated model to describe morphine pharmacokinetics in humans. J Clin Pharmacol 2019. https://doi.org/10.1002/jcph.1400 [Epub ahead of print].

[18] Lötsch J, Weiss M, Kobal G, Geisslinger G. Pharmacokinetics of morphine-6-glucuronide and its formation from morphine after intravenous administration. Clin Pharmacol Ther 1998;63(6):629−39.

[19] Crews KR, Gaedigk A, Dunnenberger HM, Klein TE, et al. Clinical Pharmacogenetics Implementation Consortium (CPIC) guidelines for codeine therapy in the context of cytochrome P450 2D6 (CYP2D6) genotype. Clin Pharmacol Ther 2012;91:321−6.

[20] West RE, Guevara MG, Mikel C, Gamez R. Detection of hydrocodone and morphine as metabolites in oral fluid by LC-MS/MS in patients prescribed codeine. Ther Drug Monit 2017;39:88−90.

[21] Oyler JM, Cone EJ, Joseph Jr RE, Huestis MA. Identification of hydrocodone in human urine following

controlled codeine administration. J Anal Toxicol 2000; 24:530—5.

[22] Yue QY, Hasselström J, Svensson JO, Säwe J. Pharmacokinetics of codeine and its metabolites in Caucasian healthy volunteers: comparisons between extensive and poor hydroxylators of debrisoquine. Br J Clin Pharmacol 1991;31:635—42.

[23] Yue QY, Svensson JO, Sjöqvist F, Säwe J. A comparison of the pharmacokinetics of codeine and its metabolites in healthy Chinese and Caucasian extensive hydroxylators of debrisoquine. Br J Clin Pharmacol 1991;31:643—7.

[24] Shah JC, Mason WD. Plasma codeine and morphine concentrations after a single oral dose of codeine phosphate. J Clin Pharmacol 1990;30:764—6.

[25] Judis J. Binding of codeine, morphine, and methadone to human serum proteins. J Pharmacol Sci 1977;66(6): 802—6.

[26] Mohammed SS, Christopher MM, Mehta P, Kedar A, et al. Increased erythrocyte and protein binding of codeine in patients with sickle cell disease. J Pharmacol Sci 1993; 82:1112—27.

[27] Chen ZR, Somogyi AA, Reynolds G, Bochner F. Disposition and metabolism of codeine after single and chronic doses in one poor and seven extensive metabolizers. Br J Clin Pharmacol 1991;31:381—90.

[28] Olkkola KT, Kontinen VK, Saari TI, Kalso EA. Does the pharmacology of oxycodone justify its increasing use as an analgesic? Trends Pharmacol Sci 2013;34:206—14.

[29] Lalovic B, Kharasch E, Hoffer C, Risler L, et al. Pharmacokinetics and pharmacodynamics of oral oxycodone in healthy human subjects: role of circulating active metabolites. Clin Pharmacol Ther 2006;79:461—79.

[30] Balyan R, Mecoli M, Venkatasubramanian R, Chidambaran V, et al. CYP2D6 pharmacogenetic and oxycodone pharmacokinetic association study in pediatric surgical patients. Pharmacogenomics 2017;18: 337—48.

[31] Kinnunen M, Piirainen P, Kokki H, Lammi P, Kokki M. Updated clinical pharmacokinetics and pharmacodynamics of oxycodone. Clin Pharmacokinet 2019;58: 705—25.

[32] Manchikanti L, Atluri S, Kaye AM, Kaye AD. Hydrocodone bitartrate for chronic pain. Drugs Today 2015;51: 415—27.

[33] Dhillon S. Hydrocodone bitartrate ER (Hysingla® ER): a review in chronic pain. Clin Drug Invest 2016;36: 969—80.

[34] Cardia L, Calapai G, Quattrone D, Mondello C, et al. Preclinical and clinical pharmacology of hydrocodone for chronic pain: a mini review. Front Pharmacol 2018;9: 1122.

[35] Toyama K, Uchida N, Ishizuka H, Sambe T, Kobayashi S. Single-dose evaluation of safety, tolerability and pharmacokinetics of newly formulated hydromorphone immediate-release and hydrophilic matrix extended-release tablets in healthy Japanese subjects without co-administration of an opioid antagonist. J Clin Pharmacol 2015;55:975—84.

[36] Felden L, Walter C, Harder S, Treede RD, et al. Comparative clinical effects of hydromorphone and morphine: a meta-analysis. Br J Anaesth 2011;107:319—28.

[37] Rosielle DA. Oral oxymorphone #181. J Palliat Med 2010;13(1):78—9.

[38] Smith HS. Clinical pharmacology of oxymorphone. Pain Med 2009;10(Suppl. 1):S3—10.

[39] Mounteney J, Giraudon I, Denissov G, Griffiths P. Fentanyls: are we missing the signs? Highly potent and on the rise in Europe. Int J Drug Pol 2015;26:626—31.

[40] Suzuki J, El-Haddad S. A review: fentanyl and non-pharmaceutical fentanyls. Drug Alcohol Depend 2017; 171:107—16.

[41] McIntyre I, Anderson DT. Postmortem fentanyl concentrations: a review. J Forensic Res 2012;3:157. https://doi.org/10.4172/2157-7145.1000157.

[42] Scholz J, Steinfath M, Schulz M. Clinical pharmacokinetics of alfentanil, fentanyl and sufentanil. An update. Clin Pharmacokinet 1996;31:275—92.

[43] Ziesenitz VC, König SK, Mahlke NS, Skopp G, et al. Pharmacokinetic interaction of intravenous fentanyl with ketoconazole. J Clin Pharmacol 2015;55(6):708—17.

[44] Woodhouse A, Mather LE. The minimum effective concentration of opioids: a revisitation with patient controlled analgesia fentanyl. Reg Anesth Pain Med 2000;25:259—67.

[45] Biedrzycki OJ, Bevan D, Lucas S. Fatal overdose due to prescription fentanyl patches in a patient with sickle cell/beta-thalassemia and acute chest syndrome: a case report and review of the literature. Am J Forensic Med Pathol 2009;30:188—90.

[46] Bodenham A, Park GR. Alfentanil infusions in patients requiring intensive care. Clin Pharmacokinet 1988;15: 216—26.

[47] Klees TM, Sheffels P, Dale O, Kharasch ED. Metabolism of alfentanil by cytochrome p4503a (cyp3a) enzymes. Drug Metab Dispos 2005;33:303—11.

[48] Pitsiu M, Wilmer A, Bodenham A, Breen D, et al. Pharmacokinetics of remifentanil and its major metabolite, remifentanil acid, in ICU patients with renal impairment. Br J Anaesth 2004;92:493—503.

[49] Wilhelm W, Kreuer S. The place for short-acting opioids: special emphasis on remifentanil. Crit Care 2008; 12(Suppl. 3):S5.

[50] Egan TD, Minto CF, Hermann DJ, Barr J, et al. Remifentanil versus alfentanil: comparative pharmacokinetics and pharmacodynamics in healthy adult male volunteers. Anesthesiology 1996;84:821—33.

[51] Clotz MA, Nahata MC. Clinical uses of fentanyl, sufentanil, and alfentanil. Clin Pharm 1991;10:581—93.

[52] Leen JLS, Juurlink DN. Carfentanil: a narrative review of its pharmacology and public health concerns. Can J Anaesth 2019;66:414—21.

[53] Kharasch ED, Greenblatt DJ. Methadone disposition: implementing lessons learned. J Clin Pharmacol 2019. https://doi.org/10.1002/jcph.1427 [Epub ahead of print].

[54] Volpe DA, Xu Y, Sahajwalla CG, Younis IR, Patel V. Methadone metabolism and drug-drug interactions: in vitro

and in vivo literature review. J Pharmacol Sci 2018;107: 2983—91.

[55] Kristensen K, Blemmer T, Angelo HR, Christrup LL, et al. Stereoselective pharmacokinetics of methadone in chronic pain patients. Ther Drug Monit 1996;18:221—7.

[56] Ahmad T, Valentovic MA, Rankin GO. Effects of cytochrome P450 single nucleotide polymorphisms on methadone metabolism and pharmacodynamics. Biochem Pharmacol 2018;153:196—204.

[57] Latta KS, Ginsberg B, Barkin RL. Meperidine: a critical review. Am J Therapeut 2002;9:53—68.

[58] Clark RF, Wei EM, Anderson PO. Meperidine: therapeutic use and toxicity. J Emerg Med 1995;13:797—802.

[59] Todd EL, Stafford DT, Bucovaz ET, Morrison JC. Pharmacokinetics of meperidine in pregnancy. Int J Gynaecol Obstet 1989;29:143—6.

[60] Beckwith MC, Fox ER, Chandramouli J. Removing meperidine from the health-system formulary-frequently asked questions. J Pain Palliat Care Pharmacother 2002;16: 45—59.

[61] Bravo L, Mico JA, Berrocoso E. Discovery and development of tramadol for the treatment of pain. Expet Opin Drug Discov 2017;12:1281—91.

[62] Miotto K, Cho AK, Khalil MA, Blanco K, et al. Trends in tramadol: pharmacology, metabolism, and misuse. Anesth Analg 2017;124:44—51.

[63] Grond S, Sablotzki A. Clinical pharmacology of tramadol. Clin Pharmacokinet 2004;43:879—923.

[64] Wu WN, McKown LA, Liao S. Metabolism of the analgesic drug ULTRAM (tramadol hydrochloride) in humans: API-MS and MS/MS characterization of metabolites. Xenobiotica 2002;32:411—25.

[65] Faria J, Barbosa J, Moreira R, Queirós O, et al. Comparative pharmacology and toxicology of tramadol and tapentadol. Eur J Pain 2018;22:827—44.

[66] Pergolizzi Jr JV, Taylor Jr R, LeQuang JA, Raffa RB, Bisney J. Tapentadol extended release in the treatment of severe chronic low back pain and osteoarthritis pain. Pain Ther 2018;7:37—57.

[67] Jain D, Basniwal PK. Tapentadol, a novel analgesic: review of recent trends in synthesis, related substances, analytical methods, pharmacodynamics and pharmacokinetics. Bull Fac Pharm Cairo Univ 2013;51: 283—9.

[68] Khanna IK, Pillarisetti S. Buprenorphine - an attractive opioid with underutilized potential in treatment of chronic pain. J Pain Res 2015;8:859—70.

[69] Gudin J, Fudin J, Nalamachu S. Levorphanol use: past, present and future. Postgrad Med 2016;128:46—53.

[70] Hayes CJ, Hudson TJ, Phillips MM, Bursac Z, et al. The influence of propoxyphene withdrawal on opioid use in veterans. Pharmacoepidemiol Drug Saf 2015;24:1180—8.

[71] Smith HS. Opioid metabolism. Mayo Clin Proc 2009;84: 613—24.

[72] Liu Z, Mortimer O, Smith CA, Wolf CR, Rane A. Evidence for a role of cytochrome P450 2D6 and 3A4 in ethylmorphine metabolism. Br J Clin Pharmacol 1995;39:77—80.

[73] Gilbert AK, Hosztafi S, Mahurter L, Pasternak GW. Pharmacological characterization of dihydromorphine, 6-acetyldihydromorphine and dihydroheroin analgesia and their differentiation from morphine. Eur J Pharmacol 2004;492:123—30.

[74] Ammon S, Hofmann U, Griese EU, Gugeler N, Mikus G. Pharmacokinetics of dihydrocodeine and its active metabolite after single and multiple oral dosing. Br J Clin Pharmacol 1999;48(3):317—22.

[75] Webb JA, Rostami-Hodjegan A, Abdul-Manap R, Hofmann U, et al. Contribution of dihydrocodeine and dihydromorphine to analgesia following dihydrocodeine administration in man: a PK-PD modelling analysis. Br J Clin Pharmacol 2001;52:35—43.

[76] Goldstein G. Pentazocine. Drug Alcohol Depend 1985; 14:313—23.

[77] Davis GA, Rudy AC, Archer SM, Wermeling DP. Pharmacokinetics of butorphanol tartrate administered from single-dose intranasal sprayer. Am J Health Syst Pharm 2004;61:261—6.

[78] Zeng Z, Lu J, Shu C, Chen Y, et al. A comparison of nalbuphine with morphine for analgesic effects and safety : meta-analysis of randomized controlled trials. Sci Rep 2015;5:10927.

[79] Davis MP, Fernandez C, Regel S, McPherson ML. Does nalbuphine have a niche in managing pain? J Opioid Manag 2018;14:143—51.

[80] Burma NE, Kwok CH, Trang T. Therapies and mechanisms of opioid withdrawal. Pain Manag 2017;7:455—9.

[81] Weaver L, Palombi L, Bastianelli KMS. Naloxone administration for opioid overdose reversal in the prehospital setting: implications for pharmacists. J Pharm Pract 2018;31:91—8.

[82] Vanky E, Hellmundt L, Bondesson U, Eksborg S, Lundeberg S. Pharmacokinetics after a single dose of naloxone administered as a nasal spray in healthy volunteers. Acta Anaesthesiol Scand 2017;61:636—40.

[83] McDonald R, Lorch U, Woodward J, Bosse B, et al. Pharmacokinetics of concentrated naloxone nasal spray for opioid overdose reversal: phase I healthy volunteer study. Addiction 2018;113:484—93.

[84] Douaihy AB, Kelly TM, Sullivan C. Medications for substance use disorders. Soc Work Publ Health 2013;28: 264—78.

[85] Alexander L, Mannion RO, Weingarten B, Fanelli RJ, Stiles GL. Development and impact of prescription opioid abuse deterrent formulation technologies. Drug Alcohol Depend 2014;138:1—6.

[86] Lee YH, Brown DL, Chen HY. Current impact and application of abuse-deterrent opioid formulations in clinical practice. Pain Physician 2017;20:E1003—23.

[87] Vosburg SK, Jones JD, Manubay JM, Ashworth JB, et al. Assessment of a formulation designed to be crush-resistant in prescription opioid abusers. Drug Alcohol Depend 2012;126:206—15.

[88] Kinzler ER, Pantaleon C, Iverson MS, Aigner S. Syringeability of morphine ARER, a novel, abuse-deterrent,

extended-release morphine formulation. Am J Drug Alcohol Abuse 2019;16:1–8.

[89] Vosburg SK, Jones JD, Manubay JM, Ashworth JB, et al. A comparison among tapentadol tamper-resistant formulations (TRF) and OxyContin® (non-TRF) in prescription opioid abusers. Addiction 2013;108:1095–106.

[90] Mustafa AA, Rajan R, Suarez JD, Alzghari SK. A review of the opioid analgesic benzhydrocodone-acetaminophen. Cureus 2018;10:e2844.

[91] Dhawan A, Modak T, Sarkar S. Transdermal buprenorphine patch: potential for role in management of opioid dependence. Asian J Psychiatr 2019;40:88–91.

[92] Goodbar NH, Hanlon KE. Implantable buprenorphine (probuphine) for maintenance treatment of opioid use disorder. Am Fam Physician 2018;97:668–70.

[93] Kaye AD, Jones MR, Kaye AM, Ripoll JG, et al. Prescription opioid abuse in chronic pain: an updated review of opioid abuse predictors and strategies to curb opioid abuse: Part 1. Pain Physician 2017;20(2S):S93–109.

[94] Keast SL, Owora A, Nesser N, Farmer K. Evaluation of abuse-deterrent or tamper-resistant opioid formulations on overall health care expenditures in a state medicaid program. J Manag Care Spec Pharm 2016;22:347–56.

CHAPTER 3

Opioid Abuse and Opioid Epidemic

INTRODUCTION

In the 1996 conference of the American Pain Society, Campbell in the presidential address suggested that "pain" should be considered with an intention to encourage doctors and nurses to listen to their patients and assess their pain. This was due to the fact that sometimes healthcare professionals ignored patients' suffering from pain. However, the intention of the presidential address was not to encourage physicians to prescribe opioids to all patients complaining about pain because only focusing on reducing pain intensity without considering the patient's overall quality of life is clearly misguided. Pain should be reduced effectively and safely. Sometimes pain experienced by a patient does not correlate with underlying disease. In such cases, a patient's pain may be magnified by psychological factors (e.g., depression, anxiety, or pain catastrophizing). If this is the case, pain interventions using drug therapy or other approaches may be ineffective. For such a patient psychological intervention or guided exercise to increase activity or even no treatment at all may at the best interest of the patient. Low back pain could be part of a "central sensitization" or fibromyalgia syndrome in which case the range of effective treatment options might be very limited. However, clinicians should be cautious in thinking that a magnetic resonance imaging scan necessarily tells us how much pain a patient feels. There are too many variables from the process of transmission of physical energy, to neural transduction, to nociceptive processing, and to the report of pain before we even get to the richness of the myriad of psychological variables [1].

Opioids may be used for treating acute pain or chronic pain. Chronic pain is defined by the International Association for the Study of Pain as "pain that persists beyond normal tissue healing time, which is assumed to be 3 months." [2]. There are dilemmas in prescribing opioids or other pain medications to a patient complaining about pain. Asking a patient to rate his or her pain intensity may have the unintended consequence of fostering the overuse of opioids leading to greater harm. However, it is also likely that some patients who might benefit from the skillful use of opioids are denied care. Sullivan and Ballantyne concluded that striving to eliminate or minimize chronic pain through opioid therapy has led to the iatrogenic injury of patients and the general population. The authors propose that the root problem may be neither the high risks nor the low efficacy of long-term opioid therapy, but rather an improper focus on reducing pain intensity. Therefore, the duty of clinicians to patients with chronic pain is not to reduce pain intensity but to improve their quality of life. It is surely not humane to provide aggressive opioid pain management aimed at reducing pain intensity if it consigns a patient to serious and potentially life-long opioid adverse effects. The idea that opioids titrated to pain intensity can reliably reduce chronic pain and improve quality of life not only exposes patients to harm but also to unrealistic and potentially damaging expectations for those in pain, and for clinicians. What matters ultimately is not whether the patient's pain intensity has reduced, but whether the patient's life has improved [3].

ANALGESIC LADDER PROPOSED BY THE WORLD HEALTH ORGANIZATION

The use of opioids for the treatment of cancer pain, as first proposed in the guidelines released in 1986 by the World Health Organization (WHO), is now supported by more than 30 years of clinical experience, and several new editions of the recommendations have been published [4,5]. The "three-step analgesic ladder," one of the central components of the guideline, has also been shown to be a safe and beneficial approach to the treatment of patients with chronic noncancer [6,7].

The WHO ladder is a stepwise approach to pain management where the first step (Step 1) recommends treating mild pain with nonopioid analgesics such as acetaminophen or nonsteroidal antiinflammatory drugs. For moderate pain relief, Step 2 of the ladder recommends using a moderate μ-opioid receptor agonist in combination with a nonopioid medication. This protocol is also effective in treating moderate postoperative pain. For more severe pain, Step 3 recommends the use of a stronger opioid and round the clock schedule

Fighting the Opioid Epidemic. https://doi.org/10.1016/B978-0-12-820075-9.00003-X

monitoring of severe pain. The WHO pain ladder lists codeine, hydrocodone, and tramadol as "weak opioids," and morphine, oxycodone, methadone, hydromorphone, and fentanyl as "strong opioids." Medications used in various steps of pain management as recommended by WHO are given in Table 3.1.

The cornerstone of the WHO document rests on five recommendations for the correct use of analgesics to make the prescribed opioid treatments effective. These recommendations include the following [8]:

1. *Oral administration of analgesics.* The oral form of medication should be administered whenever possible.
2. *Analgesics should be given at regular intervals.* To relieve pain adequately, it is necessary to prescribe the dosage to be taken at definite intervals in accordance with the patient's level of pain. The dosage of medication should be adjusted until the patient is comfortable.
3. *Analgesics should be prescribed according to pain intensity as evaluated by a scale of intensity of pain.* Analgesics should be prescribed after a thorough evaluation of pain intensity and clinical examination. The prescription must be given according to the level of the patient's pain and not according to the medical staff's perception of the pain.
4. *Dosing of pain medication should be adapted to the individual.* There is no standardized dosage in the treatment of pain. Every patient will respond differently. The correct dosage is one that will allow adequate relief of pain.
5. *Analgesics should be prescribed with a constant concern for detail.* The regularity of analgesic administration is crucial for the adequate treatment of pain. Once the distribution of medication over a day is established, it is ideal to provide a written personal program to the patient. In this way, the patient, his family, and medical staff will all have the necessary information about when and how to administer the medications.

However, over the years, the WHO pain ladder has been criticized with varying degrees of fairness. Since 1986, opioid analgesics have expanded to include new agents, fast-acting and controlled-release formulations, and fixed-dose combination products. New approaches to pain control, such as neuromodulation, nerve blocks, intrathecal drug administration, and nonpharmaceutical protocols also have been developed. Although fixed-dose combination products, such as acetaminophen and hydrocodone, as important Step 2 agents are still a useful approach for pain management, the designation of "weak" and "strong" opioids may not be as meaningful as understanding the characteristics of specific opioid agents and their appropriate doses as recommended in the original WHO protocol. In addition, a fourth step is added to the ladder to allow for interventional pain management efforts, such as peripheral nerve blockade and neurolysis, which may be appropriate for the 10%—15% of cancer patients who develop severe to very severe intractable pain. In addition, breakthrough pain should be recognized with the recommended treatment option of a fast-acting opioid analgesic. Optimal pain control is multimodal and individualized. This does not negate the value of the generalized WHO pain ladder, but clinicians should feel free to modify it, as needed, for individual patients, reflecting modern pain practice [9].

OTHER GUIDELINES IN PAIN MANAGEMENT

There are also other guidelines for the proper use of opioids in pain management. The Center for Disease Control and Prevention (CDC) recommends that chronic pain, if possible, should be treated first with nonpharmacologic therapy or with medications other than opioids. Therapy with opioids should be considered only if expected benefits are likely to outweigh risks.

TABLE 3.1
Medications Recommended for Each Step of WHO Analgesic Ladder.

Step in the Ladder	Recommendation	Example of Medications
Step 1 (for mild pain)	Nonopioid medication and adjuvant therapy	Acetaminophen Nonsteroidal antiinflammatory drugs (NSAIDs)
Step 2 (mild-to-moderate pain or pain of limited duration)	Weak opioid plus nonopioid and adjuvant therapy	Codeine/ acetaminophen Buprenorphine Tramadol Oxycodone/ acetaminophen Oxycodone/ aspirin Hydrocodone/ acetaminophen
Step 3 (moderate-to-severe pain)	Strong opioid plus nonopioid and adjuvant therapy	Morphine Levorphanol Oxycodone Hydromorphone Fentanyl Meperidine

Because of the adverse effects and risks associated with opioids, patient involvement is critical during opioid therapy. Clinicians should educate patients regarding the risk of such therapy as well as a realistic outcome. These conversations should occur at least every 3 months. According to the CDC, which reviewed studies that included one randomized controlled trial with notable limitations, opioids should initially be titrated to the lowest effective dosage. The risks and benefits of dosing with 50 morphine milligram equivalents or greater per day should be carefully evaluated, with avoidance or cautious justification of doses greater than or equal to 90 morphine milligram equivalents per day. Support for prescribing high dosages is lacking, especially because of an increased risk of overdose and known challenges associated with tapering of opioids. Increased risk of potentially fatal overdose due to reduced respiratory function has also been reported in patients who use opioids in conjunction with benzodiazepines. This class of medications should be avoided in nearly all patients receiving opioid therapy.

In addition to potentially fatal overdose, hazardous outcomes associated with opioid use include physical and psychological dependence and tolerance. Moreover, if any patient is at a higher risk of opioid use disorder the clinician should consider the criteria presented in the Diagnostic and Statistical Manual of Mental Disorders, the fifth edition to asses if a patient is suffering from any substance use disorder.

Using evidence gathered via clinical expertise, observational studies with important limitations, or randomized clinical trials with several major limitations, the CDC has concluded that treatment goals should be clearly outlined before initiating therapy with any opioid. Whether a patient's pain requires acute or long-term care, immediate-release opioids are a safer option than extended-release/long-acting opioids. The latter are more prone to abuse and have been relabeled by the US Food and Drug Administration for use in severe, long-term cases when other options, including immediate-release opioids, are ineffective or not tolerated after a 1-week trial. Opioids are hazardous in any amount and during any length of therapy. Therefore, patients should be required to return for follow-up evaluation within 1–4 weeks of receiving an initial prescription for opioids or a prescription to titrate their previously established dosage. An assessment of the harms and benefits of therapy is recommended at least every 3 months. When the harms of opioid use outweigh the benefits, physicians may consider tapering or discontinuing therapy and pursuing preferred non-opioid interventions for pain management instead.

Most states have prescription drug monitoring programs to maintain information on the amount and types of controlled substances a patient has received. When available, individual records should be reviewed by physicians before starting patients on opioid therapy, and patients' records should be rechecked throughout the course of therapy as part of routine follow-up practices. The CDC has determined that its recommendations, including all of the above, appropriate for most adult patients across evidence categories. In addition, the CDC proposes that physicians determine when and for whom urine testing is appropriate. If a physician views urine drug testing as a necessary element in the management of opioid therapy, experts contend that a reasonable approach includes screening before initiating therapy and once per year thereafter [10]. Twelve major recommendations of the CDC are summarized in Table 3.2 [11].

There are other guidelines for the use of opioids in pain management. The American Pain Society and the American Academy of Pain Medicine commissioned a systematic review of the evidence on chronic opioid therapy for chronic noncancer pain and convened a multidisciplinary expert panel to review the evidence and formulate recommendations. Although evidence is limited, the expert panel concluded that chronic opioid therapy can be an effective therapy for carefully selected and monitored patients with chronic noncancer pain. However, opioids are also associated with potentially serious harms, including opioid-related adverse effects as well as opioid dependence. The recommendations presented in this document guide patient selection and risk stratification; informed consent and opioid management plans; initiation and titration of chronic opioid therapy; use of methadone; monitoring of patients on chronic opioid therapy; dose escalations, high-dose opioid therapy, opioid rotation, and indications for discontinuation of therapy; prevention and management of opioid-related adverse effects; driving and work safety; identifying a medical home and when to obtain consultation; management of breakthrough pain; chronic opioid therapy in pregnancy; and opioid-related policies [12]. Various other guidelines have also been proposed in different countries. Detail discussion on pain management protocol is beyond the scope of this chapter.

OPIOID ABUSE AND OPIOID-RELATED FATALITIES

In the fifth edition of *Diagnostics and Statistical Manual for Mental Disorder (DSM-V)* published by the American

TABLE 3.2
CDC's Twelve Recommendations for Prescribing Opioids for Chronic Pain Outside of Active Cancer, Palliative, and End-of-Life Care.

Criteria	Recommendations
Determining when to initiate or continue opioids for chronic pain	1. Nonpharmacologic therapy and nonopioid pharmacologic therapy are preferred for chronic pain. Clinicians should consider opioid therapy only if expected benefits for both pain and function are anticipated to outweigh risks to the patient. If opioids are used, they should be combined with nonpharmacologic therapy and nonopioid pharmacologic therapy, as appropriate. 2. Before starting opioid therapy for chronic pain, clinicians should establish treatment goals with all patients, including realistic goals for pain and function, and should consider how therapy will be discontinued if benefits do not outweigh risks. Clinicians should continue opioid therapy only if there is a clinically meaningful improvement in pain and function that outweighs risks to patient safety. 3. Before starting and periodically during opioid therapy, clinicians should discuss with patients known risks and realistic benefits of opioid therapy and patient and clinician responsibilities for managing therapy
Opioid selection, dosage, duration, follow-up, and discontinuation	4. When starting opioid therapy for chronic pain, clinicians should prescribe immediate-release opioids instead of extended-release/long-acting (ER/LA) opioids. 5. When opioids are started, clinicians should prescribe the lowest effective dosage. Clinicians should use caution when prescribing opioids at any dosage, should carefully reassess evidence of individual benefits and risks when increasing dosage to 50 morphine milligram equivalents (MME)/day, and should avoid increasing dosage to $=90$ MME/day or carefully justify a decision to titrate dosage to $=90$ MME/day. 6. Long-term opioid use often begins with the treatment of acute pain. When opioids are used for acute pain, clinicians should prescribe the lowest effective dose of immediate-release opioids and should prescribe no greater quantity than needed for the expected duration of pain severe enough to require opioids. Three days or less will often be sufficient; more than 7 days will rarely be needed. 7. Clinicians should evaluate benefits and harms with patients within 1—4 weeks of starting opioid therapy for chronic pain or of dose escalation. Clinicians should evaluate the benefits and harms of continued therapy with patients every 3 months or more frequently. If benefits do not outweigh the harms of continued opioid therapy, clinicians should optimize other therapies and work with patients to taper opioids to lower dosages or to taper and discontinue opioids.
Assessing risk and addressing harms of opioid use	8. Before starting and periodically during the continuation of opioid therapy, clinicians should evaluate risk factors for opioid-related harms. Clinicians should incorporate into the management plan strategies to mitigate risk, including considering offering naloxone when factors that increase the risk for opioid overdoses, such as the history of overdose, history of a substance use disorder, higher opioid dosages ($=50$ MME/day), or concurrent benzodiazepine use, are present. 9. Clinicians should review the patient's history of controlled substance prescriptions using state prescription drug monitoring program (PDMP) data to determine whether the patient is receiving opioid dosages or dangerous combinations that put him or her at high risk for overdose. Clinicians should review PDMP data when starting opioid therapy for chronic pain and periodically during opioid therapy for chronic pain, ranging from every prescription to every 3 months. 10. When prescribing opioids for chronic pain, clinicians should use urine drug testing before starting opioid therapy and consider urine drug testing at least annually to assess for prescribed medications as well as other controlled prescription drugs and illicit drugs. 11. Clinicians should avoid prescribing opioid pain medication and benzodiazepines concurrently whenever possible. 12. Clinicians should offer or arrange evidence-based treatment (usually medication-assisted treatment with buprenorphine or methadone in combination with behavioral therapies) for patients with opioid-use disorder.

Psychiatric Association, substance abuse and substance dependence are combined into "substance use disorder." The definition of misuse and related terms are given in Table 3.3.

Over the past decade, opioid prescription drug abuse has become a national public health crisis especially among adolescents and young adults. On October 26, 2017, Acting Health and Human Services Secretary Eric D. Hargan declares the opioid crisis as a public health emergency. From 2004 to 2011, the number of prescriptions written for opioid drugs increased from 164 million to 234 million representing almost one prescription per person in the US Emergency department visits related to prescription opioids increased from 150% to 200% between 2004 and 2011 [13]. In response to this epidemic, The Drug Enforcement Administration (DEA) mandated reductions in opioid production by 25% in 2017 and 20% in 2018. The number of prescriptions for opioids declined significantly from 252 million in 2013 to 196 million in 2017 (9% annual decline over this period), falling below the number of prescriptions in 2006. In addition, data from 2017 show significant reductions in the milligram equivalence of morphine by 12.2% and in the number of patients receiving high dose opioids by 16.1% [14].

Opioids are commonly encountered in deaths related to a drug overdose. Among drug overdose deaths that mentioned at least one specific drug, the 10 most frequently mentioned drugs during 2011−16 included fentanyl, heroin, hydrocodone, methadone, morphine, oxycodone, alprazolam, diazepam, cocaine, and methamphetamine. Oxycodone ranked first in 2011, heroin during 2012−15, and fentanyl in 2016. In general, cocaine consistently ranked second or third. From 2011 through 2016, the age-adjusted rate of drug overdose deaths involving heroin more than tripled, as did the rate of drug overdose deaths involving methamphetamine. The rate of drug overdose deaths involving fentanyl and fentanyl analogs also doubled each year from 2013 through 2016, from 0.6 per 100,000 in 2013 to 1.3 in 2014, 2.6 in 2015, and 5.9 in 2016. The rate of overdose deaths involving methadone decreased from 1.4 per 100,000 in 2011 to 1.1 in 2016. In 2016, the drugs most frequently mentioned in unintentional drug overdose deaths were fentanyl, heroin, and cocaine, while the drugs most frequently mentioned in suicides by drug overdose were oxycodone, diphenhydramine, hydrocodone, and alprazolam [15].

Drug overdoses resulted in 70,237 deaths during 2017; among these, 47,600 (67.8%) involved opioids (14.9 per 100,000 population), representing a 12.0% rate increase from 2016. Synthetic opioids were involved in 59.8% of all opioid-involved overdose deaths; the rate increased by 45.2% from 2016 to 2017. From 2013 through 2017, overdose death rates increased significantly in 35 states and Washington DC. From 2016 to 2017, death rates involving cocaine and psychostimulants increased 34.4% (from 3.2 to 4.3 per 100,000) and 33.3% (from 2.4 to 3.2 per 100,000), respectively, likely contributing to increases in drug overdose deaths; however, rates remained stable for deaths involving prescription opioids (5.2 per 100,000) and heroin (4.9).

From 2016 to 2017, overdose deaths involving opioid increased among males and females and among

TABLE 3.3 Drug Use, Misuse, and Other Related Terms.	
Term	**Definition**
Proper use	Taking opioids according to prescription with follow-up with the physician as recommended.
Misuse	Use of medication without a prescription. Opioids are widely misused
Abuse	Misuse on a regular basis that may lead to addiction
Tolerance	A physiological state reached due to the regular use of a drug where a higher dosage is needed to achieve the desired effect.
Withdrawal	When opioid medication (or another habit-forming drug) is discontinued abruptly, severe unpleasant effects are observed.
Substance use disorder	A cluster of cognitive, behavioral, and physiological symptoms indicating that a person is still abusing the substance despite negative consequences. Substance abuse disorder is a psychiatric illness that should be diagnosed following the guideline of the fifth edition of *Diagnostics and Statistical Manual for Mental Disorder (DSM-V)* published by the American Psychiatric Association

persons aged ≥25 years, non-Hispanic whites (whites), non-Hispanic blacks (blacks), and Hispanics. The largest relative change occurred among blacks (25.2%), and the largest absolute rate increase was among males aged 25–44 years (an increase of 4.6 per 100,000). The largest relative change among age groups was for persons aged ≥65 years (17.2%). Death rates increased significantly in 15 states, with the largest relative changes in North Carolina (28.6%), Ohio (19.1%), and Maine (18.7%).

From 2016 to 2017, the prescription opioid-involved death rate decreased 13.2% among males aged 15–24 years but increased 10.5% among persons aged ≥65 years. These death rates remained stable from 2016 to 2017 across all racial groups and urbanization levels and in most states, although five states (Maine, Maryland, Oklahoma, Tennessee, and Washington) experienced significant decreases, and one (Illinois) had a significant increase. The largest relative changes included a 29.7% increase in Illinois and a 39.2% decrease in Maine. The highest prescription opioid-involved death rates in 2017 were in West Virginia (17.2 per 100,000), Maryland (11.5), and Utah (10.8).

Overdose deaths involving heroin declined among many groups in 2017 compared with those in 2016. The largest declines occurred among persons aged 15–24 years (15.0%), particularly males (17.5%), as well as in medium metro counties (6.1%). Rates also declined by 3.2% among whites. However, heroin-involved overdose death rates did increase among some groups; the largest relative rate increase occurred among persons aged ≥65 years (16.7%) and 55–64 years (11.6%) and among blacks (8.9%). Rates remained stable in most states, with significant decreases in five states (Maryland, Massachusetts, Minnesota, Missouri, and Ohio), and increases in three (California, Illinois, and Virginia). The largest relative decrease (31.9%) was in Ohio, and the largest relative increase (21.8%) was in Virginia. The highest heroin-involved overdose death rates in 2017 were in DC (18.0 per 100,000), West Virginia (14.9), and Connecticut (12.4).

Deaths involving synthetic opioids may be responsible for increases from 2016 to 2017 across all demographic categories. The highest death rate was in males aged 25–44 years (27.0 per 100,000), and the largest relative increases occurred among blacks (60.7%) and American Indian/Alaska Natives (58.5%). Deaths increased across all urbanization levels from 2016 to 2017. Twenty-three states and Washington DC experienced significant increases in synthetic opioid-involved overdose death rates, including eight states west of the Mississippi River. The largest relative rate increase occurred in Arizona (122.2%), followed by North Carolina (112.9%) and Oregon (90.9%). The highest synthetic opioid-involved overdose death rates in 2017 were in West Virginia (37.4 per 100,000), Ohio (32.4), and New Hampshire (30.4) [16].

WHAT CAUSED OPIOID EPIDEMIC?

Multiple societal, physiological, and psychological factors contribute to increasing opioid abuse. A majority of modern recreational opioid users begin their experience with opioids as therapeutics. Shei et al. examined possible sources of prescription opioids among patients diagnosed with opioid abuse. In this study, the authors identified commercially insured patients aged 12–64 diagnosed with opioid abuse/dependence ("abuse") through the OptumHealth Reporting and Insights medical and pharmacy claims data, 2006–12. These patients were required to have continuous eligibility over an 18-month study period surrounding the first abuse diagnosis. The authors examined whether abusers had access to prescription opioids through their prescriptions and/or to diverted prescription opioids through family members' prescriptions obtained before the abuser's first abuse diagnosis. Of the 9291 abusers meeting the selection criteria, 79.9% had an opioid prescription before their first abuse diagnosis; 20.1% of abusers did not have an opioid prescription before their first abuse diagnosis, of whom approximately half (50.8%) had a family member who had an opioid prescription before the abuser's first abuse diagnosis (compared to 42.2% of non-abusers). Similar results were found among patients initially diagnosed with opioid dependence and among abusers not previously treated with buprenorphine. The authors concluded that while most abusers had access to prescription opioids through their prescriptions, many abusers without their opioid prescriptions had access to prescription opioids through family members and may have obtained prescription opioids that way [17]. This report also suggests that the availability of opioids from a family member can be a risk factor for abuse.

Misuse of prescription refills and "doctor shopping," a situation where individuals seeking opioids may go to several different doctors to receive multiple prescriptions for the drugs, are common problems associated with prescribed opioids. Han et al. estimated that the prevalence of doctor shoppers was approximately 1.4% among those up to age 64 years and 0.5% among those 65 years and older. The gender difference in doctor shoppers among all age groups was negligible. On average, the cumulative morphine-equivalent amount of Schedule II opioid per individual obtained per year was threefold to sixfold higher for doctor shoppers than for the general population across different age and gender groups [18]. In another study, the authors included records for 146.1 million opioid prescriptions dispensed during 2008 by 76% of US retail pharmacies. Prescriptions were linked to unique patients and weighted to estimate all prescriptions and patients in the nation. Finite mixture models were used to estimate different latent patient populations having different patterns of using prescribers. On average, patients in the extreme outlying population (0.7% of purchasers), presumed to be doctor shoppers, obtained 32 opioid prescriptions from 10 different prescribers. They bought 1.9% of all opioid prescriptions, constituting 4% of weighed amounts dispensed [19].

Use of online pharmacies, some of which require little documentation, and the dark web system of encrypted websites that are designed to allow the user complete anonymity has contributed to the opioid epidemic. Early refills are a subset of prescription abuse that requires additional scrutiny by the prescriber. However, some chronic opioid users may develop tolerance requiring more medication for pain management. Katz et al. reported that the number of prescriptions, doses prescribed, and individuals receiving Schedule II prescription opioids steadily increased from 1996 to 2006, the study period. Most individuals (87.5%) used one to two prescribers, one to two pharmacies, and had no early refills in 2006. However, when a patient had more than four prescribers and more than four pharmacies to fill opioid prescriptions, questionable activity may be suspected. The authors observed questionable activities in 2748 individuals, among 47,953 prescriptions, and 2,966,056 doses in 2006. The Schedule II opioid most commonly associated with questionable activity was short-acting oxycodone. The authors concluded that primary care physicians should become a useful public health surveillance provider to monitor the medical and nonmedical use of prescription opioids and to inform public health and safety policy [20].

The US opioid epidemic has changed profoundly in the last 3 years, in ways that require substantial recalibration of the US policy response. This is due to switch from prescription opioids to heroin or even fentanyl. Kertesz has summarized the changing nature of overdose deaths in Jefferson County (home to Birmingham, Alabama) using data updated through June 30, 2016. The author observed that heroin and fentanyl dominated an escalating epidemic of lethal opioid overdose, whereas opioids commonly obtained by prescription played a minor role, accounting for no more than 15% of reported deaths in 2015. The observed changes in the opioid epidemic are particularly remarkable because they have emerged despite sustained reductions in opioid prescribing and sustained reductions in prescription opioid misuse. Among US adults, past-year prescription opioid misuse is at its lowest level since 2002. Among 12th graders, it is at its lowest level in 20 years. A credible epidemiologic account of the opioid epidemic is as follows: although opioid prescribed by physicians appears to have unleashed the epidemic before 2012, opioids prescribed by physicians no longer play a major role in opioid crisis today. The accelerating pace of the opioid epidemic in 2015−16 requires a serious reconsideration of governmental policy initiatives that continue to focus on reductions in opioid prescribing [21].

A review of surveys interviewing heroin users who used opioid pain pills before, the first time an individual used heroin range from 40% to 86% which was significant to suggest a relationship [22]. The availability of heroin in the United States is increasing. Moreover, heroin is less expensive than prescription opioids on the streets. The estimated cost of a 10 mg dose of oxycodone is approximately $10 while it is estimated 50 mg of 50% pure heroin is around the same price. Heroin use may also be favorable because of the increased potency of the drug compared to morphine; a larger amount of heroin can cross the blood−brain barrier compared to morphine [23].

Heroin Laced With Fentanyl and Fatality

In 2005, only 8.7% of opioid initiators started with heroin, but this sharply increased to 33.3% ($P < .001$) in 2015, with no evidence of stabilization. The use of commonly prescribed opioids, oxycodone, and hydrocodone, dropped from 42.4% to 42.3% of opioid initiators, respectively, to 24.1% and 27.8% in 2015, such that heroin as an initiating opioid was now more frequently endorsed than prescription opioid analgesics [24].

Heroin is often cut with fentanyl because fentanyl is cheaper than fentanyl. The wholesale cost of heroin is $65,000 per kilogram while the same amount of fentanyl costs only $3500. Illicit fentanyl is mostly produced in China and Mexico but some may be synthesized in the United States. The most illicit opioid in the United States is heroin. In 2016, heroin was eclipsed with a new threat of heroin laced with fentanyl that resulted in more mortality. Illicit fentanyl may be cut into heroin to increase profit margin and many heroin users may be unaware that they are consuming adulterated heroin containing a substantial amount of fentanyl. Fentanyl is significantly more potent than heroin and as a result risk of fatal overdose is increased in individuals abusing heroin laced with fentanyl. This is a serious issue because 91 Americans die every day from an opioid overdose but about half of them have never taken any prescription opioid [25].

WHICH OPIOIDS ARE MOST COMMONLY ABUSED?

Heroin is a Schedule I drug with no known medical use. Heroin has a very high addiction potential. Issues with heroin abuse are discussed in detail in Chapter 1. In this section, commonly encountered opioid prescription drugs that are also misused or abused are addressed.

Based on the findings of the 2015 National Survey on Drug Use and Health, Han et al. estimated that among civilian, noninstitutionalized US adults aged 18 years or older, 37.8% or an estimated 91.8 million people used prescription opioids in the prior year. In addition, an estimated 4.7% of the US population or 11.5 million people misused them, and 0.8% or 1.9 million suffered from substance abuse disorder. Among adults with prescription opioid use, the 12-month prevalence of misuse was 12.5% and the 12-month prevalence of prescription opioid use disorders was 2.1%. Among adults reporting misuse of prescription opioids in 2015, 59.9% used them without a prescription, 22.2% used them in greater amounts than directed on their prescription, 14.6% used them more often than directed, and 13.1% used them longer than directed. These categories were not mutually exclusive. Interestingly, adults aged 18−49 years had a lower prevalence of prescription opioid use than older adults. Men also showed a lower prevalence of prescription opioid use when compared with women (35.3% vs. 40.2%), and Hispanic persons had a lower prevalence than non-Hispanic white persons (31.5% vs. 40.0%) for opioid use. College graduates had a lower prevalence of prescription opioid use than adults with less than a high school education (32.1% vs. 37.3%), whereas those with some college education but without a degree had a higher prevalence (42.7% vs. 37.3%). Compared with adults with private health insurance only, uninsured adults had a lower prevalence of prescription opioid use (31.6% vs. 34.6%), whereas Medicaid beneficiaries had a higher prevalence (47.9% vs. 34.6%) of opioid use. Among adults with prescription opioid use, misuse without or with drug opioid use disorders were more commonly reported in those who had lower family incomes or were uninsured or unemployed. A major reason for opioid prescription misuse is for pain relief. A smaller percentage of people misused prescription opioids for euphoria. Among adults with prescription opioid misuse overall in 2015, 40.8% obtained prescription opioids free from friends or relatives for their most recent misuse [26]. Prescription opioid use and opioid use disorder in the US population based on the 2015 survey are summarized in Table 3.4. Interestingly opioid use disorder was highest among uninsured people and individuals 18−29 years old. Jordon et al. also reported a high prevalence of past-year prescription opioid misuse among people 11−30 years old [27].

Opioid drugs commonly prescribed in the United States include hydrocodone (e.g., Vicodin), oxycodone (e.g., OxyContin, Percocet), oxymorphone (e.g., Opana), morphine (e.g., Kadian, Avinza), codeine, fentanyl, methadone, tramadol, and others. Hydrocodone products are the most commonly prescribed for a variety of indications, including dental- and injury-related pain. Oxycodone and oxymorphone are also prescribed for moderate-to-severe pain relief. Morphine is often used before and after surgical procedures to alleviate severe pain, and codeine is typically prescribed for milder pain. However, regardless of potency, all these opioids are abused.

Butler et al. reported that some drugs are known to be widely abused, especially hydrocodone and oxycodone containing products are abused less often than prescribed volume would predict. However, other opioid drugs, such as methadone, morphine, hydromorphone, fentanyl, and oxymorphone, are abused more often than their prescribed volume would predict. Moreover, certain opioids such as hydrocodone, oxycodone, and methadone are abused through oral administration but other opioids such as morphine and hydromorphone are significantly more likely to be for abuse through intravenous injection. In general, oxycodone (both immediate-release and extended-release formulations), hydrocodone, and methadone have a higher risk of abuse than morphine, extended-release oxymorphone, and extended-release fentanyl [28].

Oxycodone Abuse

Oxycodone (14-hydroxy-7,8-dihydrocodeinone) is a strong opioid agonist that is available alone or in combination with mild analgesics. It is suitable for oral administration due to high bioavailability and may also be given intramuscularly, intravenously, subcutaneously, and rectally. However, oxycodone is not recommended for spinal administration. In analgesic potency, oxycodone is comparable to morphine.

TABLE 3.4
Opioid Use and Abuse in US Population Based on the 2015 National Survey on Drug Use and Health.

Parameter	Any Use of Opioid Prescription (%)	Proper Use of Opioid Prescription (%)	Opioid Use Disorder (%)
All ages	37.8	87.5	2.1
18–29 years	35.7	76.4	3.5
30–49 years	37.0	85.4	2.8
>50 years	39.5	93.7	1.0
Males	35.3	84.3	2.9
Females	40.2	90.1	1.4
Married	36.2	91.5	1.1
Widowed	41.0	92.4	1.6
Divorced/Separated	45.5	88.2	2.5
Never married	36.4	78.0	3.7
Family income 20,000–49,000	39.1	86.4	2.3
Family income 50,000–74,999	36.8	89.2	2.1
Family income >75,000	35.5	89.5	1.3
Private insurance	34.6	87.3	1.6
Uninsured	31.6	73.9	5.0
Medicaid	47.9	85.5	3.5

Unfortunately, the abuse potential of oxycodone is equivalent to that of morphine. When oxycodone is administered, the same precautions should be taken as with morphine or other agonist opioids [29]. Initially, oxycodone was available as an immediate-release form but later due to high abuse potential, extended-release and abuse-deterrent formulations have been introduced in the market.

Oxycodone extended-release formula was approved by the FDA in 1996 and in 1997 this drug was marketed as OxyContin. Because of the delayed absorption of oxycodone, the drug was considered to have less potential for abuse than immediate-release oxycodone. Oxy-Contin available at various dosages; 10, 15, 20, 30, 40, 60, and 80 mg. However, by November 2000, there were concerns regarding the misuse and abuse of Oxy-Contin. At the request of FDA, the manufacturer added a "Black Box Warning" to the drug label alerting patients regarding the abuse potential of this drug. Oxycodone is the first commonly used medication that received a black box warning regarding its abuse potential. Nevertheless, the drug company manufacturing OxyContin continued aggressive marketing of this drug including providing clinicians with free coupons as well as marketing videos which wrongly claimed that addiction occurs only in less than 1% patients taking OxyContin. As a result, the manufacturer, as well as pharmaceutical companies manufacturing other opioids, received warning letters from FDA cautioning them about misleading video. In 2010, a new formulation of Oxy-Contin was approved by the FDA designed to reduce its abuse potential. The tamper-resistant formula prevented the tablets from being crushed for the immediate release of all oxycodone to reach euphoria. It is important to note that other oxycodone formulations are also abused [22].

Oxycodone DETERx (Xtampza ER) is an extended-release, microsphere-in-capsule, abuse-deterrent formulation designed to retain its extended-release properties after tampering (e.g., chewing/crushing). In one study, the authors demonstrated the lower oral abuse potential of chewed and intact oxycodone DETERx than crushed immediate-release oxycodone [30].

Oxymorphone Abuse

Oxymorphone, which is structurally related to oxycodone, is also abused. Oxymorphone is a semisynthetic μ-opioid agonist, which is twice as potent as oxycodone for pain relief. Oral formulations of oxymorphone were reintroduced in the United States in 2006. Babalonis et al. examined the direct physiologic effects, relative abuse liability, analgesic profile, and overall pharmacodynamic potency of oxymorphone in comparison with identical doses of oxycodone using healthy, nondependent opioid abusers ($n = 9$) over a period of 3 weeks. Seven experimental sessions (6.5 h) were conducted, during which an oral dose of immediate-release formulations of oxymorphone (10, 20, and 40 mg), oxycodone (10, 20, and 40 mg), or placebo was administered. An array of physiologic, abuse liability and experimental pain measures was collected. At identical doses, oxymorphone produced approximately twofold less potent effects on miosis compared with oxycodone. Oxymorphone also produced a lesser magnitude of effects on measures of respiratory depression, two experimental pain models and observer-rated agonist effects. However, 40 mg of oxymorphone was similar to 40 mg of oxycodone on several abuse-related subjective ratings. Overall, oxymorphone was found to be less potent on most pharmacodynamic measures, although at higher doses, its abuse liability was similar to oxycodone [31].

Hydrocodone or Hydromorphone Abuse

Hydrocodone and hydromorphone have similar abuse potential as oxycodone. Walsh et al. using a double-blind, randomized, within-subject, placebo-controlled design examined the relative abuse potential and potency of oral oxycodone (10, 20, and 40 mg), hydrocodone (15, 30, and 45 mg), hydromorphone (10, 17.5, and 25 mg), and placebo. Healthy adult volunteers ($n = 9$) with sporadic prescription opioid abuse participated in 11 experimental sessions (6.5 h in duration) conducted in a hospital setting. All three opioids produced a typical μ-opioid agonist profile of subjective (increased ratings of liking, good effects, high and opiate symptoms), observer-rated, and physiological effects (miosis, modest respiratory depression, exophoria, and decrements in visual threshold discrimination) that were generally dose-related. Valid relative potency assays revealed that oxycodone was roughly equipotent to or slightly more potent than hydrocodone. Hydromorphone was only modestly more potent (less than twofold) than either hydrocodone or oxycodone, which was inconsistent with prior estimates arising from analgesic studies. The authors concluded that the abuse liability profile and relative potency of these three commonly used opioids do not differ substantially from one another [32].

Morphine or Codeine Abuse

Although morphine is less abused in the United States, slow-release oral morphine sulfate is one of the most abused prescription opioids in France. However, fentanyl is most commonly abused in Estonia while buprenorphine in Finland [33]. Although many abusers prefer to inject morphine, oral abuse of slow-release morphine is also gaining popularity. The FDA has currently attributed an "abuse-deterrent formulation" label to two different morphine compounds: an agonist/antagonist combination, and a morphine formulation with a physical barrier. The combination of morphine sulfate and naltrexone showed bioequivalence with extended-release morphine. Naltrexone was found in low levels or nondetectable in most of the patients after the prescription use of morphine/naltrexone combination. However, if a drug abuser tampers the capsule then orally available naltrexone would be released, causing a decreased euphoria expected after regular morphine abuse. Moreover, morphine abuse-deterrent formulation obtained with a physical barrier did not reduce drug liking compared to extended-release morphine formulation. The authors concluded that both morphine formulations may offer an alternative with lower abuse potential in the treatment of chronic pain. Nevertheless, current evidence suggests that only a little percentage of abusers may stop abusing drugs as a result of reformulation [34].

Codeine is not available as an over-the-counter medication in the United States. However, codeine is a widely used analgesic that is available for sale in pharmacies over the counter in several countries including the United Kingdom, South Africa, Ireland, France, and Australia. In these countries where codeine is available without a prescription, there have been emerging concerns about misuse of and dependence on codeine-containing combination analgesics, with increasing numbers of people presenting for help with codeine dependence at primary care and addiction treatment services [35].

Fentanyl Abuse

Heroin adulterated with fentanyl is associated with many overdose deaths as discussed in Chapters 1 and 2. In addition, fentanyl and its analogs are also abused. Fentanyl was first synthesized in 1960 by Paul Janseen in Belgium and marketed as a medicine for treating pain. It was approved by the FDA as an intravenous anesthetic in 1972, marketed under the trade name Sublimaze. Within a year of going off-patent (1981), sales of fentanyl increased 10-fold. Then, reports of misuse and illicit use by clinicians, primarily anesthesiologists and surgeons with access to the drug, were published in the medical literature. In the 1990s, fentanyl transdermal patches were introduced for widespread palliative use but the misuse of fentanyl transdermal patches also emerged. In 1994, the FDA issued a warning regarding the dangers associated with fentanyl patches, expressing that it should only be prescribed to those with severe pain that cannot be managed by less potent opioids. In 2017, the US Senate resolution called fentanyl abuse a public health crisis [36].

Following the synthesis of fentanyl in 1960, many fentanyl analogs were developed for medicinal and veterinary use including sufentanil, alfentanil, remifentanil, and carfentanil. To date, there are no reports of misuse of these pharmaceutical analogs except for carfentanil. These drugs are Schedule II drugs. However, many nonpharmaceutical fentanyl derivatives synthesized by clandestine laboratories are associated with both abuse and death. Starting in the winter of 1979, multiple opioid overdoses were identified in California from the use of "China White" or synthetic heroin, but no heroin or other known opioids were detected by toxicology analysis. The causative agent was eventually identified to be α-methylfentanyl. In California alone, fentanyl analogs were determined to be responsible for >100 overdose deaths from 1979 to 1991 [37]. Another analog, 3-methylfentanyl, emerged in 1984 in Allegheny County, Pennsylvania and was responsible for 16 fatal overdose cases [38]. α-Methyl and 3-methylfentanyl were subsequently classified as schedule I narcotics in 1981 and 1986, respectively.

In 2013, acetylfentanyl emerged as yet another fentanyl analog responsible for numerous fatalities in Rhode Island, Pennsylvania, and North Carolina. It is believed that the magnitude of this outbreak is underappreciated because acetylfentanyl is not routinely monitored by clinical and forensic toxicology laboratories [39]. Acetylfentanyl has no known medical use. As a result, in 2015, the DEA announced the scheduling of acetylfentanyl as a Schedule I drug. In 2016, the DEA declared that butyryl fentanyl and β-hydroxythiofentanyl were associated with numerous fatalities in 2015 and classified them as Schedule I narcotics. At least 40 confirmed overdose deaths involving butyryl fentanyl abuse were reported in Maryland, New York, and Oregon and at least seven confirmed overdose fatalities involving β-hydroxythiofentanyl were reported in Florida. Carfentanil, synthesized by Janseen Pharmaceutica in 1974 and used as a

general anesthetic for large animals, has also made its way into the heroin supply in the United States. The first outbreak occurred in the Midwest and Appalachian region in August—September 2016. The DEA estimated 300 carfentanil overdoses during this time. Other fentanyl analogs abused include furanylfentanyl, 4-fluorobutyrylfentanyl, 4-methoxyburtyrylfentanyl, acrylfentanyl, 4-chloroisobutyrylfentanyl, 4-fluoroisobutyrfentanyl, tetrahydrofuranfentanyl, and cyclopentylfentanyl. However, these fentanyl analogs that are also nonpharmaceuticals are more abused in Europe but less in the United States [36].

Similar to the introduction of synthetic cannabinoid and synthetic cathinone derivatives in the illicit drug market of synthetic opioids including fentanyl and its derivatives, and other chemically unrelated opioid agonists including AH-7921 and MT-45 that also emerged in the illicit drug market. Among the most frequently encountered compounds in postmortem casework are furanyl fentanyl and U-47700 (trans-3,4-dichloro-N-(2-(dimethylamino)cyclohexyl)-N-methylbenzamide). Both drugs have been reported to be present in the heroin supply and to be gaining popularity among recreational opioid users but were initially developed by pharmaceutical companies in the 1970s as candidates for development as potential analgesic therapeutic agents but never marketed [40].

Fentanyl is a full agonist at the μ-opioid receptor, and approximately 50—100 times more potent than morphine. Because fentanyl and nonpharmaceutical fentanyl derivatives are active when administered at less than 100 μg dosage, they are some of the most potent medications known to exist. Similar to other opioid prescription medications, fentanyl's effects include analgesia, anxiolysis, euphoria, drowsiness, feelings of relaxation, respiratory depression, constipation, miosis, nausea, pruritus, and cough suppression, as well as orthostatic hypotension, urinary urgency or retention, postural syncope and chest wall rigidity especially after intravenous administration. Because fentanyl and its derivatives (both pharmaceutical and nonpharmaceutical) are lipophilic, these drugs can rapidly diffuse through membranes including the blood—brain barrier. The rapid uptake of fentanyl into tissues leads to a rapid fall in serum levels, with 98.6% of the dose eliminated from plasma within 60 min, with an elimination half-life of 219 min. In patients receiving fentanyl infusions with a patient-controlled-analgesia device postoperatively, serum levels ranging from 0.3 to 0.7 ng/mL provide adequate analgesia. Hypoventilation can be appreciated at serum levels >1.5 ng/mL, which may be providing insufficient analgesia to some patients. In overdose victims, the mean serum levels may reach 25 ng/mL (range 3.0–383 ng/mL) [41]. For transdermal preparations, time to maximal concentration ranges from 27.5 to 36.8 h, with maximal serum concentrations ranging from 0.3 to 2.5 ng/mL, depending on the strength of the patch. Reports of fentanyl overdoses suggest that at least a portion of the overdoses may have occurred rapidly, possibly due to the extreme potency of these substances making it very easy to ingest a toxic quantity inadvertently. Moreover, the respiratory depressant effects of fentanyl reach their maximal levels as quickly as 2 min after injection administration. There are numerous reports of bodies being discovered with needles still attached to the arm, indicating the user died even before the needle could be removed [42]. In a study of 48 overdoses in Ohio, norfentanyl, fentanyl's metabolite, was not detected at all in 42% of the cases. This potentially suggests that the individuals died too rapidly before any fentanyl could be metabolized to any appreciable degree [43]. Given that the bioavailability and the onset of action of the intranasal route are similar to the intravenous route, the risk of overdose may also be elevated from the intranasal use of fentanyl and nonpharmaceutical fentanyl analogs [44].

Poklis et al. reported the case histories, autopsy findings and toxicology findings of two fatal intoxications involving the designer drug butyryl fentanyl which was taken orally. The quantitative analysis of butyryl fentanyl in postmortem fluids and tissues was performed by an ultrahigh performance liquid chromatography combined with tandem mass spectrometry. In the first case, butyryl fentanyl was the only drug detected with concentrations of 99 ng/mL in peripheral blood, 220 ng/mL in heart blood, 32 ng/mL in the vitreous humor, 590 ng/mL in gastric contents, 93 ng/g in the brain, 41 ng/g in the liver, 260 ng/mL in bile, and 64 ng/mL in urine. The cause of death was ruled fatal intoxication by butyryl fentanyl. In the second case, butyryl fentanyl was detected along with acetyl fentanyl, alprazolam, and ethanol. The butyryl fentanyl concentrations were 3.7 ng/mL in peripheral blood, 9.2 ng/mL in heart blood, 9.8 ng/mL in the vitreous humor, 4000 ng/mL in gastric contents, 63 ng/g in the brain, 39 ng/g in the liver, 49 ng/mL in bile, and 2 ng/mL in urine. The acetyl fentanyl concentrations were 21 ng/mL in peripheral blood, 95 ng/mL in heart blood, 68 ng/mL in the vitreous humor, 28,000 ng/mL in gastric contents, 200 ng/g in the brain, 160 ng/g in the liver, 330 ng/mL in bile, and 8 ng/mL in urine. In addition, the alprazolam concentration was 40 ng/mL and the ethanol concentration was 110 mg/dL, both measured in peripheral blood. The cause of death in the second case was ruled a mixed drug intoxication. In both cases, the manner of death was an accident [45].

O'Donnell et al. studied fentanyl-related deaths in 10 states from July to December 2016 and reported that fentanyl was detected in 56.3% of 5152 opioid overdose deaths in 10 states during that time period. Among these 2093 fentanyl positive deaths, medical examiners determined that nearly all deaths (97.1%) were related to fentanyl poisoning. Moreover, fentanyl analogs were detected in 720 opioid overdose deaths with the most common being carfentanil, (389 deaths), furanylfentanyl (182 deaths), and acetyl fentanyl (147 deaths). In addition, synthetic opioid U 47700 was responsible for 40 deaths [46].

Butler et al. reported three fatalities due to abuse of acrylfentanyl, a fentanyl analog whose presence has not yet been reported in the scientific literature in the United States. In Case 1, a 23-year-old male with a history of heroin abuse was found unresponsive in a field several feet away from his parked vehicle. His toxicology analysis showed the presence of acrylfentanyl at a concentration of concentrations of 0.3 ng/mL. In Case 2, a 43-year-old male with a history of heroin abuse was found unresponsive in his home after allegedly injecting what he thought to be heroin. Toxicology analysis confirmed acrylfentanyl concentration of 0.95 ng/mL in peripheral blood. In Case 3, a 26-year-old male with a history of heroin abuse use found unresponsive on the bathroom floor of a grocery store. Toxicological analysis of peripheral blood revealed acrylfentanyl and furanylfentanyl at concentrations of 0.32 and 0.95 ng/mL, respectively. In all three cases, the initial comprehensive blood toxicology did not reveal the presence of acrylfentanyl, highlighting the need for directed testing when scene findings and history suggest a possible substance outside the scope of traditional screening [47].

Other Abused Opioids

Tramadol is a synthetic opioid analgesic possessing a dual mechanism of action: weak agonistic effect on μ-opioid receptors and inhibition of serotonin and norepinephrine reuptake, which facilitates the monoaminergic descending inhibitory system during acute nociception and inflammatory pain. Tramadol is indicated for treating moderate-to-severe pain. It has an advantage over the other opioid analgesics, producing fewer side effects, and being less addictive. However, tramadol has similar opioid-like effects and as a result, it has the potential for abuse. Recently, mounting evidence from diversion, abuse, and overdose data suggest a growing number of tramadol abusers, particularly in developing countries [48]. Tramadol abuse has become a serious problem particularly in the Middle East, Africa, and West Asia [49]. Cognitive impairment occurs commonly among tramadol-abuse patients. Memory impairment is the most common cognitive domain to be affected [50]. Tramadol overdose can lead to fatal outcomes mostly in association with other drugs, via the potential interaction with serotonergic antidepressant medications, as well as the potential for increased central nervous system depression. However, fatality due to abuse of tramadol alone has also been reported. Gioia et al. investigated two tramadol related deaths. In Case 1, gas chromatography—mass spectrometry analysis detected tramadol in heart blood (32 µg/mL), femoral blood (23.9 µg/mL), bile (3.3 µg/mL), and urine (1.4 µg/mL). No other central nervous system depressants were detected by toxicological analysis. In Case 2, gas chromatography—mass spectrometry analysis detected tramadol in all specimens analyzed (7.5 µg/mL in the heart blood, 5.8 µg/mL in the femoral blood, and 18 µg/mL in the urine). In addition, no other central nervous system depressants were detected by toxicological analysis. The authors concluded that these deaths were due to tramadol abuse [51].

Methadone and buprenorphine are used for treating patients undergoing drug rehabilitation. In addition, both methadone and buprenorphine may also be used for pain management. However, both drugs are abused and overdose may even cause fatality. In one study over a 9-month period, the authors investigated the outcome of 13,718 patients in methadone treatment and 2716 patients in buprenorphine treatment. There were 60 sudden deaths positive for methadone (32 in-treatment) and 7 buprenorphine-positive decedents (none in treatment). Most out-of-treatment deaths occurred in people with known histories of drug misuse. Moreover, 43 methadone positive cases—19/32 in treatment, and 24/28 out-of-treatment—and 2 of the 7 buprenorphine-positive deaths were due to overdose. The risk of overdose death per 1000 people in treatment was lower for buprenorphine than for methadone [52].

COMBINED ABUSE OF OPIOIDS, OTHER DRUGS, AND ALCOHOL: A DEADLY COMBINATION

Perhaps somewhat underappreciated is the contribution of concurrent use of alcohol and other sedative agents to the mounting incidence of opioid-related morbidity and mortality, even when used appropriately. The literature suggests that benzodiazepine users are more likely to receive prescription opioids than nonbenzodiazepine users [53]. Unfortunately, benzodiazepine use contributes to later opioid use among subjects who reported no prior opioid use. In fact, benzodiazepine users are seven times more likely to have a prescription for opioid than nonbenzodiazepine users. As a result, benzodiazepines are often detected as a cointoxicant in opioid-related overdose deaths [54]. It has been estimated that the overdose rate may be 10 times higher among high dose opioid recipients (defined a taking 200 mg or more oral morphine equivalent opioid medication) who also take benzodiazepines. Another factor is that approximately 80% patients prescribed high dose opioids were also receiving prescription benzodiazepines. Benzodiazepines interact pharmacodynamically with opioids to suppress breathing. As a result, benzodiazepines increase risk with opioids if both medications are taken at bedtime [55].

Alprazolam and diazepam are two benzodiazepines commonly involved in accidental opioid deaths. In patients taking both benzodiazepines and opioids, sedative effects were observed earlier and for a longer duration when opioids and benzodiazepines were given together as opposed to separately, possibly representing a drug—drug interaction. Based on postmortem blood concentrations, one case report suggested that oxycodone's metabolism was reduced by concomitant clonazepam intake thus significantly increasing the toxicity of oxycodone [56].

A growing body of research indicates a link between chronic pain and excess alcohol consumption. Up to one-quarter of people seeking treatment for chronic pain report excess alcohol use and more than two-fifths of problem-drinkers report experiencing moderate-to-severe pain in the past month [57]. According to the 2008 National Survey on Drug Use and Health, heavy drinkers (defined as five or more drinks on the same occasion on each of 5 days in the last 30 days) are more likely to use prescription opioids for nonmedical purposes than are lighter drinkers or nondrinkers. Based on a study of 15 light drinkers (eight males) and 14 moderate drinkers (eight males) who received 0, 10, and 20 mg of oxycodone orally, the authors observed that alcohol-drinking status did not modulate the abuse liability-related effects of oxycodone [58]. However, ethanol together with oxycodone causes greater ventilatory depression than either alone, the magnitude of which is clinically relevant. Elderly patients may be more affected than younger patients [59].

One of the major concerns when combining alcohol with opioid analgesics is the pharmacokinetic consequence of "dose dumping," which is defined as the unintended, rapid release (over a short period of time) of the entire amount or a significant fraction of the drug contained in a modified-release dosage form. Alcohol is linked to dose-dumping effects across specific extended-release opioid formulations, and it significantly increases their dangers, as well as their abuse liability. For example, in vitro studies of Avinza (Pfizer Inc; morphine sulfate extended-release capsules) displayed accelerated release of morphine in the presence of alcohol. Box warnings for Avinza, as well as other extended-release/long-acting opioids, advise patients not to drink alcoholic beverages or use prescription or nonprescription medications containing alcohol during therapy, as it may result in the rapid release and absorption of a potentially fatal dose of opioid [53].

High dosage buprenorphine (Subutex()) has been prescribed as replacement therapy for major opioid dependencies but when the drug is combined with benzodiazepines, alcohol or other central nervous system depressants, it may cause severe overdose or even fatality. Ferrant et al. reported three fatal buprenorphine-related poisonings after snorting buprenorphine but these three subjects also showed a significant amount of alcohol in the blood. The blood concentrations of buprenorphine for the three cases (between 6.1 and 15.4 ng/mL) remained within the limits of therapeutic concentrations. In addition, alcohol concentrations varied between 106 and 182 mg/dL, which were significantly lower than lethal blood alcohol concentrations. Therefore, the cause of death was probably due to respiratory depression caused by a combination of alcohol and buprenorphine. The authors concluded that fatal poisoning cases involving snorting buprenorphine may cause death, even with moderate consumption of alcohol. It is probably due to a high buprenorphine bioavailability with this route [60].

Prevalence estimates for alcohol use disorder in opioid-dependent patients vary depending on the study, but approximately one-third of the patients in methadone treatment are assumed to have alcohol problems. An Irish study estimated the prevalence of problem alcohol use among patients attending primary care for methadone treatment at 35%. Data from the British National Treatment Outcome Research Study suggest that almost half of the patients in residential programs drink alcohol and just over a third of those in community programs drink above the recommended levels. A Swiss 2-year longitudinal study found occasional alcohol abuse in 38%–47% of methadone patients and daily abuse in 20%–24%. In a large German study in 1685 heroin users and patients on opioid maintenance treatment (with methadone or codeine), 28% of participants consumed more than 40 g alcohol/day. The average alcohol consumption was significantly higher in heroin users than in methadone-treated patients. Predictors of alcohol use were male sex, daily cannabis, and benzodiazepine consumption, and longer duration of drug use. Meta-analyses of US clinical trials demonstrated alcohol abuse disorder in 38% and 45% of patients seeking treatment for opioid or stimulant use, respectively. Alcohol use disorder is associated with an increased risk of fatal overdose in opioid dependence and numerous studies have identified alcohol abuse or dependence as a risk factor for mortality in opioid-dependent patients [61–63].

Heroin is a Schedule I drug with no known medical use. Heroin is highly abused in many heroin fatalities alcohol is also detected as a co-toxicant. In fact, alcohol can reduce the toxic concentration of heroin significantly. As a result, severe toxicity and even fatality may occur after consuming much less heroin in comparison to people who may abuse heroin without consuming alcohol. Sutlović and Definis-Gojanović reported three deaths due to fatal combinations of heroin and alcohol. The first case of poisoning was related to a young couple, a 30-year-old man and a 28-year-old woman who was found dead in a car, surrounded by cans of a variety of alcoholic drinks. Two needles were found beside the bodies as well. The victims were registered drug abusers who had been in withdrawal programs. The second case was a 29-year-old man who was found dead in a house. Three fresh injection marks were visible on his right arm, and two needles were near his body. He was not known as a drug addict but attempted to commit suicide recently. Carboxyhemoglobin along with heroin, meconin, acetaminophen, 6-acetylmorphine, codeine, noscapine, and papaverine was detected in the blood. Ethanol, being a respiratory depressant, combined with morphine drastically increases the risk of rapid death due to respiration failure [64]. In a study based in New Mexico, the authors observed that, during 1990–2005, the 196% increase in single drug category overdose death was driven by prescription opioids alone and heroin alone; the 148% increase in multidrug category overdose death was driven by heroin/alcohol and heroin/cocaine. Hispanic males had the highest overdose death rate, followed by white males, white females, Hispanic females, and American Indians. The most common categories causing death were heroin alone and heroin/alcohol among Hispanic males, heroin/alcohol among American Indian males, and prescription opioids alone among white males and all female subpopulations [65].

PRESCRIPTION MONITORING

The United States is enduring a devastating opioid misuse epidemic leading to over 33,000 deaths per year from both prescription and illegal opioids where almost half of these deaths are attributable to prescription opioids. As a result, both Federal and state governments have introduced opioid prescription guidelines as well as opioid prescription monitoring to combat this opioid misuse epidemic. In 2016, the Centers for Disease Control and Prevention released their Guidelines for Prescribing Opioids for Chronic Pain with a goal to control and regulate the prescription of opioids by clinicians. In addition, the Food and Drug Administration (FDA), the Drug Enforcement Administration (DEA), and the Department of Justice have also participated in controlling this epidemic. The DEA working with the Department of Justice has enforcement power to prosecute pill mills and physicians for illegal prescribing. The DEA could also implement use of prescription drug monitoring programs (PDMPs), currently administered at the state level, and use of electronic prescribing for schedule II and III medications. The FDA has the authority to approve new and safer formulations of immediate- and long-acting opioid medications as well as abuse-deterrent opioid formulation. More importantly, the FDA can also ask pharmaceutical companies to cease the manufacturing of certain opioid formulation with high abuse potential. Additionally, state agencies play a critical role in reducing overdose deaths, protecting public safety, and promoting the medically appropriate treatment of pain. One of the states' primary roles is the regulation of the practice of medicine and the insurance industry within their borders. Utilizing this authority, states can both educate physicians about the dangers of opioids and make physician licensure dependent on registering and using prescription drug monitoring programs when prescribing controlled substances including opioids. Almost every state has implemented such a program to some degree. Further, states have the flexibility to promote innovative interventions to reduce harm such as legislation allowing naloxone access without a prescription. Although relatively new, these types of laws have allowed first responders, patients, and families to have access to a lifesaving drug. Finally, states are at the forefront of litigation against pharmaceutical manufacturers. This approach is described as similar to the initial steps in fighting tobacco companies. In addition

to fighting for dollars to support drug treatment programs and education efforts, states are pursuing these lawsuits as a means of holding pharmaceutical companies accountable for misleading marketing of a dangerous product [66].

Prescription drug monitoring programs are statewide databases that gather information from pharmacies on dispensed prescriptions of controlled substances and, as such, are promising tools to help combat the prescription opioid epidemic. Prescribers, pharmacists, law enforcement agencies, and medical licensure boards are among the typical users of these databases. Prescription drug monitoring programs implemented since the late 1990s are all electronic instead of paper-based; typically allow users, especially prescribers, access by means of an online portal; and cover a wider range of controlled substances, compared to drug monitoring programs implemented earlier However, prescription drug monitoring programs dated back to the late 1930s. A new wave of implementation began in the early 2000s, and all states except Missouri have either implemented or upgraded their prescription drug monitoring programs or have enacted legislation to do so. Effective prescription drug monitoring programs can help change prescriber behavior by identifying patients at a high risk of doctor shopping or diversion. They also allow law enforcement agencies and medical licensure boards to monitor aberrant prescribing practices. A prescription drug monitoring program was associated with more than a 30% reduction in the rate of prescribing of Schedule II opioids. This reduction was seen immediately following the launch of the program and was maintained in the second and third years afterward. Increased utilization of these programs and the adoption of new policies and practices governing their use may have contributed to sustained effectiveness [66].

PDMPs are a principal strategy used in the United States to address prescription drug abuse. Moyo et al. compared opioid use pre- and post-PDMP implementation and estimated differences of PDMP impact by reason for Medicare eligibility and plan type. The authors analyzed opioid prescription claims in US states that implemented PDMPs relative to non-PDMP states during 2007–12. These states include Florida, Louisiana, Nebraska, New Jersey, Vermont, Georgia, Wisconsin, Maryland, New Hampshire, and Arkansas. Based on a study of 310,105 disabled and older adult Medicare enrollees, the authors concluded that PDMPs are associated with reductions in opioid use, measured by volume, among disabled and older adult Medicare beneficiaries in the United States compared with states that do not have PDMPs [67].

Currently, 49 out of 50 states in the United States have implemented PDMP. Although the core function of PDMP in most states is clinical monitoring, there is growing interest in using PDMP data for public health surveillance, epidemiological, and health service-related research. One advantage of PDMO is that it collects data of all relevant prescription fills regardless of payers. As a result, supplementing pharmacy claims with PDMP data has the potential to identify out of pocket cash payments. It has been speculated that out of pocket payment for prescription opioids is associated with misuse and abuse of opioids. In one study based on 33,592 Medicaid beneficiaries who filled a total of 555,103 opioid prescriptions, the authors observed that 13.5% of these prescriptions could not be matched with Medicaid claims indicating out of pocket purchases. Hydromorphone (30%), fentanyl (18%), and methadone (15%) were the most likely to lack a matching claim. The three largest predictors for missing claims were opioid fills that overlapped with other opioids, long-acting opioids, and fills at multiple pharmacies. The authors concluded that opioid prescription fills that are out of pocket are associated with high-risk opioid use [68].

CONCLUSIONS

Pain is considered as the fifth vital sign and proper pain management is essential for the well-being of patients. Moderate-to-severe pain is often treated with opioids but opioids also carry a risk of misuse and abuse. In addition, heroin is widely abused and is associated with many fatalities. Consuming excessive alcohol during opioid therapy is associated with an increased risk of adverse effects. In many fatalities, both opioid and alcohol could be detected during postmortem analysis. More recent development and marketing of abuse-deterrent opioid formulation may reduce abuse liabilities of certain opioids. PDMPs have been adopted by 49 states and such an approach is useful in combating the current opioid epidemic.

REFERENCES

[1] Campbell JN. The fifth vital sign revisited. Pain 2016; 157(1):3−4.

[2] International Association for the Study of Pain. Classification of chronic pain descriptions of chronic pain syndromes and definitions of pain terms. Prepared by the International Association for the Study of Pain, Subcommittee on Taxonomy. Pain 1986;(Suppl. l): S1−226.

[3] Sullivan MD, Ballantyne JC. Must we reduce pain intensity to treat chronic pain? Pain 2016;157(1):65−9.

[4] World Health Organization. Cancer and palliative care: report of WHO expert committee. World Health Organization Technical Report Series 804. Geneva, Switzerland. 1990.

[5] World Health Organization. Cancer pain relief and palliative care. With a guide to opioid availability. Geneva, Switzerland: World Health Organization; 1996.

[6] Vargas-Schaffer G. Patient therapeutic education: placing the patient at the centre of the WHO analgesic ladder. Can Fam Physician 2014;60:235−41.

[7] Miller E. The World Health Organization analgesic ladder. J Midwifery Womens Health 2004;49:542−5.

[8] Vargas-Schaffer G. Is the WHO analgesic ladder still valid: twenty-four years of experience. Can Fam Physician 2010; 56:514−7.

[9] Raffa RB, Pergolizzi Jr JV. A modern analgesics pain 'pyramid'. J Clin Pharm Therapeut 2014;39:4−6.

[10] Bredemeyer M. CDC develops guideline for opioid prescribing. Am Fam Physician 2016;93:1042−3.

11. Dowell D, Haegerich TM, Chou R. CDC guideline for prescribing opioids for chronic pain — United States, 2016. MMWR Recomm Rep 2016;65(No. RR-1):1−49. https:// doi.org/10.15585/mmwr.rr6501e1.

[12] Chou R, Fanciullo GJ, Fine PG, Adler JA, et al. Clinical guidelines for the use of chronic opioid therapy in chronic noncancer pain. J Pain 2009;10:113−30.

[13] Soelberg CD, Brown Jr RE, Du Vivier D, Meyer JE, Ramachandran BK. The US opioid crisis: current federal and state legal issues. Anesth Analg 2017;125:1675−81.

[14] Manchikanti L, Sanapati J, Benyamin RM, Atluri S, et al. Reframing the prevention strategies of the opioid crisis: focusing on prescription opioids, fentanyl, and heroin epidemic. Pain Physician 2018;21:309−26.

[15] Hedegaard H, Bastian BA, Trinidad JP, Spencer M, Warner M. Drugs most frequently involved in drug overdose deaths: United States, 2011−2016. Natl Vital Stat Rep 2018;67:1−14.

[16] Scholl L, Seth P, Kariisa M, Wilson N, Baldwin G. Drug and opioid-involved overdose deaths — United States, 2013−2017. Morb Mortal Wkly Rep 2019;67(5152): 1419−27.

[17] Shei A, Rice JB, Kirson NY, Bodnar K, et al. Sources of prescription opioids among diagnosed opioid abusers. Curr Med Res Opin 2015;31(4):779−84.

[18] Han H, Kass PH, Wilsey BL, Li CS. Increasing trends in Schedule II opioid use and doctor shopping during 1999−2007 in California. Pharmacoepidemiol Drug Saf 2014;23:26−35.

[19] McDonald DC, Carlson KE. Estimating the prevalence of opioid diversion by "doctor shoppers" in the United States. PLoS One 2013;8(7):e69241.

[20] Katz N, Panas L, Kim M, Audet AD, et al. Usefulness of prescription monitoring programs for surveillance–analysis of Schedule II opioid prescription data in Massachusetts, 1996−2006. Pharmacoepidemiol Drug Saf 2010; 19:115−23.

[21] Kertesz SG. Turning the tide or riptide? The changing opioid epidemic. Subst Abuse 2017;38:3−8.

[22] Kanouse AB, Compton P. The epidemic of prescription opioid abuse, the subsequent rising prevalence of heroin use, and the federal response. J Pain Palliat Care Pharmacother 2015;29:102−14.

[23] Schaefer CP, Tome ME, Davis TP. The opioid epidemic: a central role for the blood brain barrier in opioid analgesia and abuse. Fluids Barriers CNS 2017;14(1):32.

[24] Cicero TJ, Ellis MS, Kasper ZA. Increased use of heroin as an initiating opioid of abuse. Addict Behav 2017;74: 63−6.

[25] Pergolizzi Jr JV, LeQuang JA, Taylor Jr R, Raffa RB, NEMA Research Group. Going beyond prescription pain relievers to understand the opioid epidemic: the role of illicit fentanyl, new psychoactive substances, and street heroin. Postgrad Med 2018;130:1−8.

[26] Han B, Compton WM, Blanco C, Crane E, et al. Prescription opioid use, misuse, and use disorders in U.S. Adults: 2015 national survey on drug use and health. Ann Intern Med 2017;167:293−301.

[27] Jordan AE, Blackburn NA, Des Jarlais D, Hagan H. Past-year prevalence of prescription opioid misuse among those 11 to 30 years of age in the United States: a systematic review and meta-analysis. J Subst Abuse Treat 2017; 77:31−7.

[28] Butler SF, Black RA, Cassidy TA, Dailey TM, Budman SH. Abuse risks and routes of administration of different prescription opioid compounds and formulations. Harm Reduct J 2011;8:29.

[29] Pöyhiä R, Vainio A, Kalso E. A review of oxycodone's clinical pharmacokinetics and pharmacodynamics. J Pain Symptom Manag 1993;8:63−7.

[30] Kopecky EA, Fleming AB, Levy-Cooperman N, O'Connor M, Sellers EM. Oral human abuse potential of oxycodone DETERx® (Xtampza® ER). J Clin Pharmacol 2017;57:500−12.

[31] Babalonis S, Lofwall MR, Nuzzo PA, Walsh SL. Pharmacodynamic effects of oral oxymorphone: abuse liability, analgesic profile and direct physiologic effects in humans. Addiction Biol 2016;21:146−58.

[32] Walsh SL, Nuzzo PA, Lofwall MR, Holtman JR. The relative abuse liability of oral oxycodone, hydrocodone and hydromorphone assessed in prescription opioid abusers. Drug Alcohol Depend 2008;98:191−202.

[33] Peyriere H, Nogue E, Eiden C, Frauger E, et al. Evidence of slow-release morphine sulfate abuse and diversion: epidemiological approaches in a French administrative area. Fundam Clin Pharmacol 2016;30:466−75.

[34] Fanelli A, Sorella MC, Ghisi D. Morphine sulfate abuse-deterrent formulations for the treatment of chronic pain. Expet Rev Clin Pharmacol 2018;11:1157−62.

[35] Nielsen S, Van Hout MC. Over-the-Counter codeine-from therapeutic use to dependence, and the grey areas in between. Curr Top Behav Neurosci 2017;34:59−75.

[36] Armenian P, Vo KT, Barr-Walker J, Lynch KL. Fentanyl, fentanyl analogs and novel synthetic opioids: a comprehensive review. Neuropharmacology 2018;134(Pt A): 121−32.

[37] Henderson GL. Fentanyl-related deaths: demographics, circumstances, and toxicology of 112 cases. J Forensic Sci 1991;36:422−33.

[38] Hibbs J, Perper J, Winek CL. An outbreak of designer drug−related deaths in Pennsylvania. J Am Med Assoc 1991;265:1011−3.

[39] Rogers JS, Rehrer SJ, Hoot NR. Acetylfentanyl: an emerging drug of abuse. J Emerg Med 2015;50:433−6.

[40] Mohr AL, Friscia M, Papsun D, Kacinko SL, et al. Analysis of novel synthetic opioids U-47700, U-50488 and furanyl fentanyl by LC-MS/MS in postmortem casework. J Anal Toxicol 2016;40:709−17.

[41] Martin TL, Woodall KL, McLellan BA. Fentanyl-related deaths in Ontario, Canada: toxicological findings and circumstances of death in 112 cases (2002−2004). J Anal Toxicol 2006;30:603−10.

[42] Suzuki J, El-Haddad S. A review: fentanyl and non-pharmaceutical fentanyls. Drug Alcohol Depend 2017; 171:107−16.

[43] Burns G, DeRienz RT, Baker DD, Casavant M, Spiller HA. Could chest wall rigidity be a factor in rapid death from illicit fentanyl abuse? Clin Toxicol (Phila) 2016;54(5): 420−3.

[44] Foster D, Upton R, Christrup L, Popper L. Pharmacokinetics and pharmacodynamics of intranasal versus intravenous fentanyl in patients with pain after oral surgery. Ann Pharmacother 2008;42:1380−7.

[45] Poklis J, Poklis A, Wolf C, Hathaway C, et al. Two fatal intoxications involving butyryl fentanyl. J Anal Toxicol 2016;40:703−8.

[46] O'Donnell JK, Halpin J, Mattson CL, Goldberger BA, Gladden RM. Deaths involving fentanyl, fentanyl analogs, and U-47700 - 10 states, July−December 2016. MMWR Morb Mortal Wkly Rep 2017;66(43):1197−202.

[47] Butler DC, Shanks K, Behonick GS, Smith D, et al. Three cases of fatal acrylfentanyl toxicity in the United States and a review of literature. J Anal Toxicol 2018;42:e6−11.

[48] Verri P, Rustichelli C, Palazzoli F, Vandelli D, et al. Tramadol chronic abuse: an evidence from hair analysis by LC tandem MS. J Pharmaceut Biomed Anal 2015;102: 450−8.

[49] Abdel-Hamid IA, Andersson KE, Waldinger MD, Anis TH. Tramadol abuse and sexual function. Sex Med Rev 2016; 4:235−46.

[50] Bassiony MM, Youssef UM, Hassan MS, Salah El-Deen GM, et al. Cognitive impairment and tramadol dependence. J Clin Psychopharmacol 2017;37:61−6.

[51] Gioia S, Lancia M, Bacci M, Suadoni F. Two fatal intoxications due to tramadol alone: autopsy case reports and review of the literature. Am J Forensic Med Pathol 2017;38: 345−8.

[52] Bell JR, Butler B, Lawrance A, Batey R, Salmelainen P. Comparing overdose mortality associated with methadone and buprenorphine treatment. Drug Alcohol Depend 2009;104:73−7.

[53] Gudin JA, Mogali S, Jones JD, Comer SD. Risks, management, and monitoring of combination opioid,

benzodiazepines, and/or alcohol use. Postgrad Med 2013;125:115−30.

[54] Webster LR. Considering the risks of benzodiazepines and opioids together. Pain Med 2010;11:801−2.

[55] Webster LR, Reisfield GM, Dasgupta N. Eight principles for safer opioid prescribing and cautions with benzodiazepines. Postgrad Med 2015;127:27−32.

[56] Burrows DL, Hagardorn AN, Harlan GC, Wallen ED, Ferslew KE. A fatal drug interaction between oxycodone and clonazepam. J Forensic Sci 2003;48:683−6.

[57] Larance B, Campbell G, Peacock A, Nielsen S, et al. Pain, alcohol use disorders and risky patterns of drinking among people with chronic non-cancer pain receiving long-term opioid therapy. Drug Alcohol Depend 2016; 162:79−87.

[58] Zacny JP, Drum M. Psychopharmacological effects of oxycodone in healthy volunteers: roles of alcohol-drinking status and sex. Drug Alcohol Depend 2010; 107:209−14.

[59] van der Schrier R, Roozekrans M, Olofsen E, Aarts L, et al. Influence of ethanol on oxycodone-induced respiratory depression: a dose-escalating study in young and Elderly individuals. Anesthesiology 2017;126:534−42.

[60] Ferrant O, Papin F, Clin B, Lacroix C, et al. Fatal poisoning due to snorting buprenorphine and alcohol consumption. Forensic Sci Int 2011;204:e8−11.

[61] Soyka M. Alcohol use disorders in opioid maintenance therapy: prevalence, clinical correlates and treatment. Eur Addiction Res 2015;21:78−87.

[62] Darke S, Zador D. Fatal heroin 'overdose': a review. Addiction 1996;91:1765−72.

[63] Nolan S, Klimas J, Wood E. Alcohol use in opioid agonist treatment. Addiction Sci Clin Pract 2016;11:17.

[64] Sutlović D, Definis-Gojanović M. Fatal poisoning by alcohol and heroin. Arh Hig Rada Toksikol 2007;58(3):323−8.

[65] Shah NG, Lathrop SL, Reichard RR, Landen MG. Unintentional drug overdose death trends in New Mexico, USA, 1990−2005: combinations of heroin, cocaine, prescription opioids and alcohol. Addiction 2008;103:126−36.

[66] Bao Y, Pan Y, Taylor A, Radakrishnan S, et al. Prescription drug monitoring programs are associated with sustained reductions in opioid prescribing by physicians. Health Aff (Millwood) 2016;35:1045−51.

[67] Moyo P, Simoni-Wastila L, Griffin BA, Onukwugha E, et al. Impact of prescription drug monitoring programs (PDMPs) on opioid utilization among Medicare beneficiaries in 10 US States. Addiction 2017;112:1784−96.

[68] Hartung DM, Ahmed SM, Middleton L, Van Otterloo J, et al. Using prescription monitoring program data to characterize out-of-pocket payments for opioid prescriptions in a state Medicaid program. Pharmacoepidemiol Drug Saf 2017;26:1053−60.

CHAPTER 4

Genetic Factors Associated With Opioid Therapy and Opioid Addiction

INTRODUCTION

Many opioids are metabolized via CYP2D6 enzyme, which is subjected to much genetic polymorphism. For certain drugs such as codeine, it must be converted into morphine by the action of CYP2D6 for pharmacological action because most analgesic effect of codeine is due to it metabolite morphine. In addition, polymorphisms of genes encoding opioid receptors are associated with not only success of pain management therapy using opioids but also may increase vulnerability of some individuals to opioid abuse. In addition, polymorphisms of genes encoding other neurotransmitter systems may increase susceptibility of certain individuals for drug or alcohol abuse. However, no single gene that is directly linked to drug and or alcohol abuse has been identified. Therefore, unlike many genetic disorders such as sickle cell anemia or cystic fibrosis which are monogenic disorder, drug or alcohol abuse is not a monogenetic disorder. Instead both genetic factors and environmental factors are associated with increased risk of alcohol or drug abuse in certain individuals.

More recently, in guidelines published by American Psychiatry Association, both alcohol and drug abuse have been classified under one psychiatric disorder known as substance use disorder (SUD). The Diagnostic and Statistical Manual of Mental Disorders 5 (DSM-5) defines SUD as a constellation of behaviors involved in compulsive drug seeking, including impaired control of substance use, impaired social interactions with others because of substance use, risky drug use (e.g., substance use in hazardous settings), and pharmacological changes (e.g., experiencing withdrawal symptoms). Further, the DSM-5 defines addiction as the most severe, chronic stage of the SUD diagnosis, which is characterized by substantial loss of self-control, manifesting in compulsive drug-seeking behavior despite the desire to discontinue use [1].

SUBSTANCE ABUSE DISORDER: ROLE OF GENE AND ENVIRONMENT

Alcohol and drug addictions are psychiatric disorders associated with maladaptive behaviors where a person has persistent, compulsive, and uncontrolled use of alcohol or a drug. Both genetic makeup of a person and environmental factors, as well as complex interactions between gene and environment, may increase vulnerability of a person to substance abuse disorder. The human genome contains 3.2 billion nucleotides of DNA [2]. Most of the genetic polymorphisms are SNP (single-nucleotide polymorphism), and it has been estimated that more than 11 million SNPs occur in human genome with a frequency of more than 1% (if a polymorphism occurs at a frequency less than 1%, it is considered rare but when occurs at a frequency of more than 1%, it is called polymorphic). Although SNPs are the most common genetic variation, other variations including deletion, insertion, and duplication also occur.

A genetic variation results in altered expression and function of a protein which may be linked to reward from alcohol or drug abuse. Family, twin, and linkage study can provide information regarding a potential association between a phenotype and addiction. Genome-wide linkage study can identify location of the genome that is associated with the particular phenotype. Moreover, through case-control candidate gene association studies, as well as contribution of a genetic polymorphism in making an individual more susceptible to alcohol or drug abuse, can be identified. In addition, polymorphisms in genes that encode various enzymes responsible for alcohol and drug metabolism contribute to toxicity, as well as alcohol/drug addiction.

Family studies indicated that children of parents with high-risk alcohol dependence, or who are from families where one member is diagnosed with a SUD, are at much greater risk for developing alcohol problems. Consequently, family studies have demonstrated SUD clusters within families, implicating a role for both genetic and environmental influences. There are substantial genetic influences on drug use disorder that are expected to influence multiple neurotransmission pathways, and these influences are particularly important within the dopaminergic system. Genetic influences involved in other aspects of SUD etiology, including drug processing and metabolism, are also identified. Studies of gene-environment interaction emphasize the importance of environmental context in SUD. In addition, epigenetic studies indicate drug-specific changes in gene expression, as well as differences in gene expression, related to the use of multiple substances. Further, gene expression is expected to differ by a stage of SUD such as substance initiation versus chronic substance use. It has been assumed that genetic factors contribute approximately 50%, while environmental factors are responsible for approximately 50% in making people more susceptible to drug or alcohol abuse, including opioid abuse [3]. However, various studies have also indicated that heritability range of drug addiction may vary widely depending on the particular drug, for example, heritability may account for 39% in case of addiction for hallucinogens, but heritability factor may be as high as 72% in case of cocaine addiction [4].

In general, it is assumed that SUD is an organic brain disease caused by cumulative effects of alcohol and or drug on various neurotransmitters. Some drugs mimic neurotransmitters, for example, opioids, while other abused drugs alter neurotransmission by interacting with molecular components of one or more neurotransmitter systems. More than 180 neurotransmitters have been described. The central function of the human brain is due to presence of 100 billion neurons, and everything humans do relies on is the communication between neurons where various neurotransmitters play important roles. Common neurotransmitters are acetylcholine, serotonin, dopamine, GABA (gamma-aminobutyric acid), epinephrine, norepinephrine, and glutamate. Neurotransmitters also can be divided under two broad groups: inhibitory or excitatory. Excitatory neurotransmitters stimulate neurons and the brain, while inhibitory neurotransmitters have calming effect, but some neurotransmitters may play dual role. The monoaminergic neurotransmitter systems constitute the primary reward pathway in the human brain. Monoamines can be further divided into catecholamines (dopamine epinephrine and norepinephrine) and serotonin. Dopamine is synthesized from tyrosine that plays an important role in regulating emotional and motivational behavior through the mesolimbic dopaminergic pathway that also includes rewards after consuming food and also reward related to administration of drugs of abuse. Dopamine neurons are located in the ventral tegmental area (VTA) of the brain and projecting into the nucleus accumbens (NAc). After receiving the appropriate signal, dopamine is released by presynaptic neuron into the synapse where dopamine binds with dopamine receptor. The unbound dopamine is then taken up by dopamine transporters (DAT) for future utilization or degradation. The mesolimbic dopaminergic pathway is the major reward pathway of the brain. As a result, dysfunctions in the dopaminergic systems are involved in several pathological conditions, including Parkinson's disease, Tourette's syndrome, drug addiction, and hyperactivity disorders. In addition to dopamine neurotransmitter system, other neurotransmitter systems are also involved in substance abuse disorder as mutation of certain genes controlling receptors of transmitter in the system is associated with increased susceptibility of SUD. However, multiple genetic mutations are responsible for overall effects.

Opiates and opioids are metabolized by various isoenzymes of cytochrome P450 (CYP) mixed function drug-metabolizing enzymes. As expected, polymorphisms of such enzymes especially polymorphisms of CYP2D6 play a major role in efficacy of a particular opioid medication in pain control in individual patients. Currently, effects of various polymorphisms of drug metabolism enzymes on efficacy or toxicity of opioid therapy are clearly understood. However, effects of genetic polymorphism of genes encoding various receptors and transmitters in the brain have not been well documented in the literature, except polymorphisms of genes encoding opioid receptors in the brain. In this chapter, focus is on polymorphisms of genes encoding drug metabolizing enzymes and genetic polymorphisms of genes encoding opioid receptors. However, other neurotransmitters playing some role in opioid addiction will also be discussed briefly.

DRUG-METABOLIZING ENZYMES

The liver is the major organ primarily responsible for biotransforming endogenous compounds and exogenous lipophilic chemicals into water-soluble substances that are more easily eliminated from the body. However, enzymes present in the gut can also metabolize certain drugs. However, drugs may be metabolized by the liver enzymes in two different steps: Phase I and Phase II. Phase I reactions involve oxidation, reduction, or hydrolysis. Phase II biosynthetic conjugation reactions include glucuronidation, sulfation, acetylation, methylation, glutathione conjugation, or amino acid conjugation to endogenous substrates or metabolites formed during Phase I metabolism. Collectively, both Phase I and Phase II reactions significantly increase polarity and hydrophilicity of metabolites resulting in enhanced renal elimination from the body. However, certain drug after Phase I metabolism is excreted in bile and may undergo enterohepatic circulation. For example, mycophenolic acid is conjugated with glucuronic acid during Phase II metabolism forming mycophenolic acid glucuronide, which is transported into bile and is hydrolyzed by gut bacteria into mycophenolic acid. As a result, a second peak is observed after oral administration of mycophenolic acid. Major enzymes involved in Phase I reaction are isoenzymes of CYP mixed function oxidase enzymes. Other enzymes involved in Phase I reaction include flavin containing monooxygenases, monoamine oxidase, cyclooxygenase, dihydrodiol dehydrogenase, quinone oxidoreductase, alcohol dehydrogenase, and aldehyde dehydrogenase. Many endogenous compounds are also metabolized by these enzymes. Enzymes responsible for Phase I metabolism are listed in Table 4.1, while enzymes responsible for Phase II metabolism are listed in Table 4.2.

In addition to enzymes responsible for drug metabolism, drug transporters also play important role in pharmacology of certain drugs. The best-characterized drug transporter is the multidrug resistance protein MDR1 (also called P-glycoprotein). MDR1 is a glycosylated membrane-bound protein expressed mainly in intestines, liver, kidneys, and brain. A large number of structurally unrelated drugs are substrates for MDR1, which regulates their intestinal absorption, hepatobiliary secretion, renal secretion, and brain transport. Metabolic pathways of major opioids are listed in Table 4.3.

Phenotype demonstrating variation in individual response to certain drugs was first recognized in early 1950s where antimalarial drugs were found to cause in vivo hemolysis in patients with glucose 6-phosphate deficiency. Later, the pharmacogenetic differences in a number of Phase I and Phase II enzymes have been reported explaining many interindividual responses to metabolism of various drugs. Polymorphisms of genes encoding various enzymes, including CYP isoenzymes, alcohol dehydrogenase, aldehyde dehydrogenase, dihydropyrimidine dehydrogenase, and esterases, involved in Phase I metabolism have been well documented in the literature. However, not all CYP isoenzymes are polymorphic in nature. In addition, polymorphism of

TABLE 4.1
Enzymes Involved in Phase I Metabolism.

Enzyme	Comments
CYP3A4	CYP3A4 is the most abundant CYP (approximately 30%) and responsible for metabolism of almost 50% of all drugs in the market. Fentanyl is metabolized by CYP3A4.
CYP2D6	Representing only 2% of CYP enzymes, approximately 25% drugs are metabolized by this enzyme, including many opioids. This enzyme is very polymorphic and plays most important role in metabolism of codeine, hydrocodone, oxycodone, and tramadol.
CYP2B6	A minor CYP enzyme metabolizing approximately over 60 drugs, including methadone.
CYP2C9	Metabolizes warfarin, phenytoin, and other drugs. This enzyme is very polymorphic.
CYP1A1, CYP1A2, CYP1B1, CYP2C8, CYP2E1, and CYP2C19	Minor CYP enzymes metabolizing various drugs, some of which are polymorphic.
Monooxygenase enzymes	This enzyme incorporates one hydroxyl group into substrates in many metabolic pathways (including endogenous compounds).
Alcohol dehydrogenase	Converts ethanol (and related alcohols) into acetaldehyde.
Alcohol dehydrogenase	Converts acetaldehyde into acetate and also metabolizes other aldehydes.

TABLE 4.2
Enzymes Involved in Phase II Metabolism.

Enzymes	Comments
Uridine diphosphate glucuronosyltransferase (UDP-glucuronosyltransferase)	Conjugates glucuronic acid to a Phase I metabolite or a parent drug.
Sulfotransferase	The soluble sulfotransferases catalyze sulfation of molecules and are expressed mainly in the liver. Sometimes this enzyme adds sulfate to same molecule, where UDP-glucuronosyltransferase adds glucuronic acid.
N-Acetyl transferase	Transfers an acetyl group to a drugs molecule, for example, conversion of procainamide to N-acetyl procainamide.
Methyl transferase	Transfers a methyl group.

genes encoding Phase II enzymes such as uridine diphosphate glucuronosyltransferase (UGT), N-acetyl transferase, and others have also been reported. However, various CYP isoenzymes play important role in metabolism of various abused drugs. Therefore, only polymorphisms of genes encoding CYP isoform are discussed in detail in the chapter.

CYP FAMILY OF ENZYMES

The CYP family of enzymes has many isoenzymes that play very important role in the Phase I metabolism of many drugs. The CYP proteins comprise a superfamily of heme-containing proteins found in many organisms, with over 7700 known members across all species studied. The human CYP superfamily of enzymes is encoded by 57 functional genes and 58 pseudogenes producing over 50 isoenzymes (three families) in human. The name of the isoenzymes is derived from the characteristic maximum spectral absorption at 450 nm. While CYP3A4 is the most abundant enzyme (30%), other enzymes such as CYP1A2 (13%), CYP2A6 (4%), CYP2C (20%), CYP2D6 (2%), and CYP2E1 (7%) are also important members of CYP family of isoenzymes. In general, CYP3A4 is responsible for metabolism of approximately 50% of drugs, while CYP1A2, CYP2A6,

TABLE 4.3
Metabolic Pathways of Major Opioids.

Drug	Phase I Enzyme	Phase I Metabolite	Phase II Enzyme	Phase II Metabolite
Codeine	CYP2D6 CYP3A4	Morphine (active) Norcodeine (inactive)	UDP-glucuronosyl-transferase	Morphine-3-glucuronide (inactive) Morphine-6-glucuronide (active)
Morphine	None	None	UDP-glucuronosyl-transferase	Morphine-3-glucuronide (inactive) Morphine-6-glucuronide (active)
Oxycodone	CYP2D6 CYP3A4	Oxymorphone Noroxycodone	UDP-glucuronosyl-transferase	Inactive metabolites
Hydrocodone	CYP2D6 CYP3A4	Hydromorphone Norhydrocodone	UDP-glucuronosyl-transferase	Hydromorphone-6-glucuronide Hydromorphone-3-glucuronide
Tramadol	CYP2D6	O-desmethyl tramadol		
Methadone	CYP3A4, CYP1A2, CYP2C19, CYP2C19, CYP2B6	EEDP (2-ethylidene, 1, 5-dimethyl-3, 3-diphenylpyrrolidine) (inactive)		
Fentanyl	CYP3A4	Norfentanyl		

CYP2C, CYP2D6, and CYP2E1 are responsible for 13%, 4%, 25%, 2%, and 7% metabolism of all drugs. Overall, CYP families of enzymes are responsible for metabolism of approximately 80% of all drugs [5]. Some of the genes encoding CYP isoenzymes are highly polymorphic, while others show less polymorphisms or without important functional polymorphisms with respect to drug metabolism. While CYP isoenzymes encoded by CYP2D6, CYP2C9, CYP2C19, CYP2A6, and CYP2B6 genes are highly polymorphic, other CYP isoenzymes such as CYP1A1, CYP1A2, CYP2E1, and CYP3A4 do not show widely different metabolic activities due to functional polymorphism [6].

Polymorphisms of genes encoding CYP isoenzymes may result in wide interindividual variation in drug metabolism. Based on metabolism, individuals are classified into four groups:

- Extensive metabolizers (EMs): These individuals belong to "normal population" and have two functional alleles. As a result, these individuals metabolize drug at expected normal rate.
- Poor metabolizers (PMs): These individuals carry two defective alleles and as a result have enzymes with poor or no activity. These individuals metabolize a particular drug slowly or not at all.
- Intermediate metabolizers (IMs): These individuals are heterozygous for a defective allele or carrying two alleles encoding enzymes with reduced activity. These individuals metabolize a particular drug at a rate in between EM and PM.
- Ultrarapid metabolizers (UMs): These individuals carry more than two active gene copies in the same allele and as a result produce enzyme with much higher activity than normal individuals.

Polymorphism of CYP2D6

Despite low hepatic content (2%), CYP2D6 enzyme is very important in drug metabolism because approximately 25% of all drugs are metabolized by this enzyme. The CYP2D6 gene is located on the long arm of chromosome 22 (22q13.1). Both SNP and copy number variation (CNV) have been observed in this gene, and over 80 different alleles have been described. The most important polymorphisms include CYP2D6*2, CYP2D6*3, CYP2D6*4, CYP2D6*5, CYP2D6*6, CYP2D6*10, CYP2D6*17, and CYP2D6*41 [7]. Over 90% PMs are attributable to three different defective mutations: CYP2D6*5 (total deletion of gene), CYP2D6*3 (adenine deletion in exon 5: rs35742686), and CYP2D6*4 (adenine to guanine transition at the junction of intron 3 and exon 4: rs389207). CYP2D6*17 allele found commonly in black population also encodes enzyme

with reduced activity. The CNV in CYP2D6 gene includes deletion, duplications, and multiplications up to 13 genes in tandem. The UMs have more than one active gene on one allele (CYP2D6*2xN duplication). Many opioids are metabolized by CYP2D6. As a result, polymorphism of CYP2D6 gene plays a major role in effective pain management when opioids metabolized by CYP2D6 are prescribed.

Polymorphism of CYP2B6

Cytochrome P450 2B6 (CYP2B6) is a minor isoform of CYP mixed-function oxidase family of drug-metabolizing enzymes. Expression is highly variable both between individuals and within individuals, owing to nongenetic factors, genetic polymorphisms, inducibility, and irreversible inhibition by many compounds. Drugs metabolized mainly by CYP2B6 include over 60 drugs, including artemisinin, bupropion, cyclophosphamide, efavirenz, ketamine, and methadone. The CYP2B6 is one of the most polymorphic CYP genes in humans, and variants have been shown to affect transcriptional regulation, splicing, mRNA and protein expression, and catalytic activity. Some variants appear to affect several functional levels simultaneously, thus, combined in haplotypes, leading to complex interactions between substrate-dependent and substrate-independent mechanisms. The most common functionally deficient allele is CYP2B6*6, which occurs at frequencies of 15 to over 60% in different populations. The allele leads to lower expression in liver due to erroneous splicing resulting in an enzyme with much lower activity compared to normal CYP2B6 enzymatic activity. Therefore, individuals who are carrier of this genetic mutation are PMs of drugs that are metabolized by CYP2B6. Another important variant of CYP2B6 gene is CYP2B6*18, which is predominantly found in Africans (4%−12%) and does not express functional CYP2B6 enzyme. As a result, people with CYP2B6*18 allele are also PMs. In contrast, CYP2B6*4 allele encodes enzyme with higher activity than the enzyme coded by the wild-type gene (CYP2B6*1). Therefore, individuals with CYP2B6*4 are UMs. To date, at least 20 allelic variants and some subvariants (*1B through *29) have been described, and some alleles are associated with clinically significant differences in drug metabolism and drug response. Moreover, a large number of uncharacterized variants are currently emerging from different ethnicities in the course of the 1000 Genomes Project. The CYP2B6 gene polymorphism is clinically relevant not only for treatment with some opioids (methadone, bupropion) but also for HIV-infected patients treated with the reverse transcriptase inhibitor efavirenz [6].

Polymorphism of CYP2C9

CYP2C9 found predominately in the liver is 92% homologous with CYP2C19 but has different substrate specificity. Human CYP2C9 accounts for approximately 20% of total hepatic CYP content and metabolizes approximately 15% clinically used drugs, including S-warfarin, tolbutamide, phenytoin, losartan, diclofenac, and celecoxib. The wild type is *CYP2C9*1*, which is the normal gene encoding CYP2C9 enzyme with normal enzymatic activity. More than 33 polymorphisms of the gene encoding CYP2C9 enzyme have been reported (*1B through to *34), but most common variants are *CYP2C9*2* (5.5-fold decreased activity) and *CYP2C9*3* (27-fold decreased activity) alleles that produce enzyme with significantly reduced activity. Homozygous for *CYP2C9*3* are PMs. Other alleles with reduced activities include *CYP2C9*4*, *CYP2C9*5*, and *CYP2C9*30*.

*CYP2C9*2* and *CYP2C9*3* differ from the wild-type *CYP2C9*1* by a single-point mutation: *CYP2C9*2* is characterized by a 430C > T exchange in exon 3, resulting in an Arg144Cys amino acid substitution, whereas *CYP2C9*3* shows an exchange of 1075A > C in exon 7, causing an Ile359Leu substitution in the catalytic site of the enzyme. *CYP2C9*2* is frequent among Caucasians with approximately 1% of the population being homozygous carriers and 22% heterozygous. The corresponding figures for the *CYP2C9*3* allele are 0.4% and 15%, respectively. Worldwide, a number of other variants have also to be reported. The *CYP2C9* gene polymorphisms are relevant for the efficacy and adverse effects of numerous nonsteroidal antiinflammatory agents, sulfonylurea antidiabetic drugs, and, most critically, oral anticoagulants belonging to the class of vitamin K epoxide reductase inhibitors. Numerous clinical studies have shown that the *CYP2C9* gene polymorphism should be considered in warfarin therapy, and practical algorithms on how to consider it in therapy are available. These studies have highlighted the importance of the *CYP2C9*2* and *3* alleles. Warfarin has served as a practical example of how pharmacogenetics can be utilized to achieve maximum efficacy and minimum toxicity [7].

Polymorphisms of Other CYP Isoenzymes

The CYP2C19 isoenzyme plays an important role in the efficacy and safe use of many drugs. The fields of medicine where clinical outcome particularly depends on CYP2C19 polymorphism are gastroenterology, cardiology, psychiatry, mycology, and oncology. CYP2C19 is involved in proton pump inhibitors metabolism, thus it can influence reflux therapy, ulcer prevention, and *Helicobacter pylori* eradication treatment. For CYP2C19,

2%—5% Caucasian population and 20% Asian population are PMs. There are over 25 different alleles of CYP2C19 gene, but in Asian population, *CYP2C19*2* and *CYP2C19*3* allele together account for 100% of defective alleles. In Caucasian, 85% of PMs are homozygous for *CYP2C19*2*. In general, *CYP2C19*2*—*8* represent inactive version of enzyme, but *CYP2C19*17* allele encodes enzyme with increased activity than the enzyme encoded by the wild-type gene [4].

The CYP2C19 enzyme also plays a vital role in the two bioactivation steps of clopidogrel leading to lower (*CYP2C19*17* carriers) or higher (*CYP2C19*2* carriers) risk of major adverse cardiovascular events. It affects the antidepressant treatment and methadone replacement therapy, as well as voriconazole prophylaxis. The presence of a *2 allele is associated with longer relapse-free time or better survival and the *17 allele with more favorable outcomes in breast cancer patients treated with tamoxifen [8].

CYP3A isoenzymes are the predominant subfamily of CYP enzymes, making it one of the most important drug-metabolizing enzymes. The genes encoding CYP enzymes are expressed primarily in the liver and small intestine. The CYP3A activities are sum of the activities of four isoenzymes: CYP3A4, CYP3A5, CYP3A7, and CYP3A43. Genes encoding CYP3A4, CYP3A5, CYP3A7, and CYP3A43 enzyme are located on a cluster on chromosome 7. At the enzyme level, CYP3A4 is the major isoform followed by CYP3A5, CYP3A43, and CYP3A7. Many polymorphisms of *CYP3A4* gene have been described, but such polymorphisms have little effects on drug metabolism because only few polymorphisms results in altered CYP3A4 enzymatic activities. Overall variation in *CYP3A4* genotype contributes only to minor extent or only in rare cases to the interindividual variation in CYP3A4 enzymatic activity. At present, *CYP3A4*22* allele appears to be the most relevant genetic variation (0.04—0.08 frequency) that encodes an enzyme with reduced activity. Currently, there are two alleles (*CYP3A4*20* and *CYP3A4*26*) that lead to complete loss of function of the CYP3A4 enzyme. However, these alleles are reported only in two case reports [9]. Shi et al. reported that genetic polymorphism of *CYP3A4* gene (*CYP3A4*1B*) may have impact on dose requirement of tacrolimus [10]. However, for other drugs, no clear association between this allele and drug metabolism has been observed. More research is needed to clarify effects (if any) of polymorphisms of *CYP3A4* gene on interindividual differences in drug metabolism.

CYP3A4 and CYP3A5 enzymes share approximately 85% sequence homology and overlapping substrate specificities. Polymorphism of *CYP3A5* genes may lead to nonfunctioning enzymes with three alleles *CYP3A5*3*, *CYP3A5*6*, and *CYP3A5*7* encode nonfunctional enzymes. The most frequent is *CYP3A5*3* with a frequency ranging from 0.12 to 0.35 in Africans and 0.88 to 0.97 in Caucasians [9].

POLYMORPHISMS OF CYP2D6 AND OPIOID THERAPY

Polymorphisms of *CYP2D6* gene play very important role in the metabolism of certain opioids. As a result, success or failure in pain management with a particular opioid which is substrate for CYP2D6 depends on polymorphism of the gene encoding this enzyme. Codeine is a prodrug that is metabolized into morphine by CYP2D6. Therefore, a person who is a PM of CYP2D6 may not get adequate pain relief when treated with codeine. In contrast, an UM may experience toxicity due to accumulation of morphine in the blood. In addition to conversion into morphine, codeine is also converted into norcodeine by CYP3A4. Moreover, morphine is conjugated with glucuronic acid into morphine-3-glucuronide and morphine-6-glucuronide. Other opioids metabolized by CYP2D6 include hydrocodone, oxycodone, and tramadol. In contrast, morphine is not only subjected to Phase I metabolism but is also converted into morphine-3-glucuronide and morphine-6-glucuronide by UDP-glucuronosyltransferase.

CYP2D6 Gene Polymorphism and Codeine Therapy

CYP2D6 gene polymorphisms have a profound effect on therapy with codeine. CYP2D6 activity scores relate to phenotype classification system. In general, each allele is assigned an activity score with a 0 score for a nonfunctional allele, 0.5 for a reduced function allele, and 1.0 for a fully functional alleles, for example, *1. The total CYP2D6 activity score is the sum of value assigned to each allele. The typical score varies from 0 to 3, but in rare cases, the value may exceed 3. For example, in an individual carrying wild-type gene in both alleles such as *CYP2D6*1*1*, the sum of activity score should be $1 + 1 = 2$ and the person is an EM who metabolizes codeine normally as expected in majority of the population. If the score exceeds 3, then the individual is UMs. Likely, phenotypes based on CYP2D6 allele activity scores are summarized in Table 4.4. However, subjects with genotypes giving rise to an activity score of 1.0 which can be due to a

TABLE 4.4
Likely Phenotype Based on CYP2D6 Allele Activity Scores.

Phenotype	Activity Score	Example of Diplotypes (Activity Score)
Extensive metabolizers (77%–92%)	1.0–2.0	CYP2D6*1/*1 (2.0) CYP2D6*1/*2 (2.0) CYP2D6*2/*2 (2.0) CYP2D6*1/*41 (1.5) CYP2D6*1/*10 (1.5) CYP2D6*1/*3 (1.0) CYP2D6*1/*4 (1.0) CYP2D6*1/*41 (1.0)
Intermediate metabolizers (2%–11%)	0.5	CYP2D6*4/*10 (0.5) CYP2D6*5/*41 (0.5) CYP2D6*3/*41 (0.5)
Poor metabolizers (5%–10%)	0	CYP2D6*4/*4 (0) CYP2D6*4/*5 (0) CYP2D6*4/*6 (0) CYP2D6*5/*5 (0)
Ultrarapid metabolizers (5%–10%)	>2.0–3.0 (it may even exceed 3 but rarely)	CYP2D6*1/*1 × 2 (3.0) CYP2D6*/*2 × 2 (3.0)

diplotype containing one functional and one nonfunctional allele or containing two reduced function allele $(0.5 + 0.5 = 1.0$ activity) are considered by some investigators as IMs. Regardless of term used, these individuals (extensive or IMs) have lower CYP2D6 activity compared with people with activity score of 2.0 or higher activity than classical IMs with an activity score of 0.5 [11]. Activity scores are listed in Table 4.4.

The analgesic property of codeine is due to its metabolite morphine and morphine-6-glucuronide because codeine has a 200-fold weaker affinity of μ-opioid receptors compared to morphine [12]. The association of CYP2D6 metabolizer phenotype with the formation of morphine is well documented. Studies have shown a decrease in morphine formation and decrease of analgesic effects in PMs compared with EMs. In contrast, much higher morphine levels are expected in UMs. Linares et al. studied CYP2D6 phenotype-specific codeine population pharmacokinetics and reported that a codeine pharmacokinetic pathway model developed by the authors accurately fitted the time courses of plasma codeine and its metabolites morphine, morphine-6-glucuronide, and morphine-3-glucuronide. The population model indicated that about 10% of a codeine dose was converted to morphine in poor metabolizer phenotype subjects. In contrast, approximately 40% of a codeine dose was converted to morphine in EMs, and about 51% was converted to morphine in UMs. The population model further indicated that only about 4% of morphine formed from codeine was converted to morphine-6-glucuronide, the active metabolite in poor metabolizer phenotype subjects, but about 39% of the morphine formed from codeine was converted to morphine-6-glucuronide in extensive metabolizer phenotypes. Interestingly, approximately 58% of morphine was converted into morphine-6-glucuronide in UMs [13].

In one study, a single dose of 30 mg codeine was administered to 12 UMs of CYP2D6 substrates carrying CYP2D6 gene duplication, 11 EMs, and three subjects with phenotype of PM. Genotyping was performed using polymerase chain reaction-restriction fragment length polymorphism methods and a single base primer extension method for characterization of the gene duplication alleles. Pharmacokinetics was measured over 24 h after drug intake, and codeine and its metabolites in plasma and urine were analyzed by liquid chromatography with tandem mass spectrometry (LC-MS/MS). The authors observed significantly higher area under the curve for morphine in plasma in UMs compared with EMs. In general, CYP2D6 genotypes predicting ultrarapid metabolism resulted in about 50% higher plasma concentrations of morphine and its glucuronides compared with the EMs. The authors concluded that it might be good if physicians would know about the CYP2D6 gene duplication genotype of their patients before administering codeine [14].

VanderVaart et al. studied genotype of 45 mothers who received codeine for controlling postpartum pain. In this group, 14 patients were EMs, 26 IMs, 3 UMs (CYP2D6*1/*2 × N, CYP2D6*2/*17 × N, and CYP2D6*2/*2 × N), and 2 PMs (CYP2D6*4/*4 and CYP2D6*4/*5). The authors reported that two patients who were PMs reported no analgesia from receiving codeine, whereas two out of three UMs reported immediate pain relief after taking codeine but discontinued codeine due to dizziness and constipation [15].

Significant toxicity and even fatality in UMs after administration of codeine has been reported. An individual who has duplicated or amplified active CYP2D6 genes is considered to have ultrarapid

metabolism. Approximately 7—10% of Caucasian's are poor CYP2D6 metabolizers, whereas 1 to 7% of Caucasians and more than 25% of Ethiopians have gene duplications and are classified as having ultrarapid metabolism. Gasche et al. described severe toxicity in a 62-year-old man who was admitted to the hospital with severe pneumonia. On day 4, he experience severe respiratory distress and went into coma. At the time of the patient's coma and respiratory depression, his plasma morphine level was significantly elevated to 80 ng/mL (expected range, 1—4 ng/mL), the morphine-3-glucuronide level was 580 ng/mL (expected range, 8—70 ng/mL), and the morphine-6-glucuronide level was 136 ng/mL (expected range, 1—13 ng/mL). Among these metabolites, only morphine and morphine-6-glucuronide have clinically relevant opioid activity. CYP2D6 genotyping showed three or more functional alleles, a finding consistent with ultrarapid metabolism. Intravenous administration of naloxone (0.4 mg) that was repeated two times resulted in a dramatic improvement in the patient's level of consciousness and the patient recovered completely after two more days. The authors concluded that life-threatening toxicity in this patient after administration of standard dose of codeine was due to ultrarapid phenotype [16].

There is a case report of fatality in a 2-year-old child who was an UM of codeine. The boy with snoring and sleep apnea underwent elective adenotonsillectomy and was sent home with instructions for 10—12.5 mg of codeine and 120 mg of acetaminophen syrup to be administered 4—6 h for pain control. On second day, postsurgery fever and wheezing developed, and next morning, at 9 a.m., the child died. Postmortem analysis of femoral blood using gas chromatography/mass spectrometry confirmed the presence of codeine and morphine (32 ng/mL). No other drug or metabolite was detected. Respiratory depression with serum morphine level exceeding 20 ng/mL has been documented in young children. Therefore, elevated toxic morphine level was responsible for respiratory arrest and death of this boy. CYP genotyping revealed functional duplication of CYP2D6 allele resulting in the UM phenotype [17].

Kelly et al. reported three cases where two infants died after receiving codeine while one infant survived after severe toxicity. All patients received codeine after adenotonsillectomy for obstructive sleep apnea syndrome. A 4-year-old boy showed postmortem morphine level of 17.6 ng/mL, which was highly elevated based on the dosage. The boy was CYP2D6 enzyme UM (CYP2D6*1/*2 × N). Another 5-year-old boy also died, who showed blood codeine concentration of 79 ng/mL but highly elevated morphine level of 30 ng/mL in postmortem blood. It was highly likely that the child was an UM. A 3-year-old girl showed severe toxicity after receiving codeine postoperatively but survived. Her blood morphine level was 17 ng/mL. She was genotyped as EM (CYP2D6*1/*1), but the authors commented that EM genotype often overlaps with UM genotype [18].

Virbalas et al. reported that ethnic distribution in their study subjects paralleled recent local census data. A total of 154 children (80.6%) had a haplotype that corresponds to extensive codeine metabolism, 18 children (9.42%) were identified as UMs, and 16 children (8.37%) were IMs. Only 3 children in their cohort (1.57%) were PMs. Patients identifying as Caucasian or Hispanic had an elevated incidence of being UM (11.3% and 11.2%, respectively) with extensive variability within subpopulations. The authors concluded that the clinically significant effect of UMs reinforces safety concerns regarding the use of codeine and related opiates. A patient-targeted approach using pharmacogenomics may mitigate adverse effects by individualizing the selection and dosing of these analgesics [19].

Death of an infant from breast milk of the mother who was an UM and also receiving codeine postpartum has also been reported, indicating serious safety issue of codeine in UMs. A newborn male infant born after an unremarkable pregnancy and delivery (birth weight 3.88 kg, 90th percentile) developed difficulty breastfeeding and increasing lethargy at 7 days of age. At 11 days of age, he was taken to a pediatrician owing to concerns about his skin color and decreased milk intake. Subsequently, on day 13, an ambulance team found the baby cyanotic and without vital signs. Resuscitation, which was initiated at home and continued in the hospital's emergency department, was unsuccessful. Full postmortem analysis failed to identify an anatomic cause of death. Postmortem toxicologic testing using gas chromatography—mass spectrometry revealed a blood concentration of morphine at 70 ng/mL (expected level 10—12 ng/mL) and acetaminophen at 5.9 µg/mL. Review of the medical records revealed that, in the immediate postpartum period, the mother was prescribed Tylenol 3 (codeine 30 mg and acetaminophen 500 mg). Initially she took two tablets twice daily, but she halved the dose on postpartum day 2 owing to somnolence and constipation. Following the development of poor neonatal feeding, the mother

expressed milk and stored it in a freezer. Analysis of the milk for morphine revealed a concentration of 87 ng/mL. Genotype analysis indicated that the mother was heterozygous for a CYP2D6*2 allele and a CYP 2D6*2x2 gene duplication. In essence, the mother had 3 functional CYP2D6 genes and would be classified as an UM. Both the father and the infant possessed two functional CYP2D6 alleles (CYP2D6*1/*2 genotypes). The authors concluded that high amount of morphine in the breast milk due to UM phenotype of the mother causes death in that infant [20].

Current recommendation is that codeine should not be used in children, obese adolescents (between 12 and 18 years old), or those with obstructive sleep apnea or lung disease. In addition, codeine should not be prescribed to UMs of CYP2D6 phenotype due to risk of severe toxicity and even fatality. Codeine should not be used in PM due to inadequate analgesic effect. IM of CYP2D6 should avoid codeine after a failed initial attempt of pain control. However, majority of the population (approximately 80%) are EM of CYP2D6 phenotype who metabolizes codeine normally. Codeine can be prescribed to these people in recommended dosage for pain control [21].

CYP2D6 Gene Polymorphism and Tramadol Therapy

Tramadol is metabolized by CYP2D6 to pharmacologically active O-desmethyltramadol (200 times more affinity than tramadol for opiate receptors). Tramadol is also metabolized by CYP2B6 and CYP3A4 enzymes into N-dimethyl tramadol (inactive metabolite). Lassen et al. based on a systematic review of 56 studies, involving tramadol and pharmacogenomics, concluded that only the effects of CYP2D6 gene polymorphisms on metabolism of tramadol and consequent effect of pain relief has been thoroughly studied, and results show clinical significance [22]. Clinical studies have shown that PMs often fail to exhibit analgesic effect after administration of tramadol compared with EMs. Based on a study of 300 patients who received tramadol postoperatively, Stamer et al. observed that percentage of nonresponders was significantly higher in PM group (46.7%) compared with EM group (21.6%). Moreover, patients who were PMs also consumed more tramadol. The authors concluded that PMs showed a lower response rate to postoperative tramadol analgesia [23].

The CYP2D6*10 allele is a common allele with a frequency ranging from 51.3% to 70% in Chinese population. This allele encodes CYP2D6 enzyme with significantly reduced enzymatic activity. Wang et al.

based on a study of gastric cancer patients recovering from gastrectomy observed that homozygous for CYP2D6*10 allele (n = 20) required significantly more tramadol for adequate pain control in 48 h period compared with patients who were heterozygous for CYP2D6*10 (n = 26) or patients who did not carry CYP2D6*10 allele (n = 17). Therefore, PMs required more tramadol for adequate pain control [24]. Susce described a case of 85-year-old woman who was a PM and had a history of problems with opioid analgesic, including codeine, oxycodone, and tramadol, which are all metabolized by CYP2D6 enzyme. However, when that genetic information was transmitted to the physician in her subsequent visit to the hospital, she showed better response to hydrocodone. The authors concluded that CYP2D6 genotype is clinically useful in opioid pain management [25]. Based on these studies, it appears that tramadol has reduced clinical efficacy in PMs.

Pharmacokinetic studies have shown higher plasma concentration of O-desmethyltramadol in UMs after administration of tramadol compared with EMs. As a result, UMs not only experience greater analgesia but also show higher incidences of side effects and toxicity from tramadol. A 5.5-year-old boy (21 kg) was admitted to the hospital for adenotonsillectomy to treat obstructive sleep apnea. He had an uneventful 6-hour perioperative stay and was discharged home. Approximately 8 h after arriving home, he received one oral dose of tramadol of 20 mg. The following morning, he was found to be lethargic and was taken to the emergency department. Upon arrival, the patient had a pediatric Glasgow coma scale score of 8, pinpoint pupils, minimal respiratory effort, frequent apneic events, and an oxygen saturation of 48% on room air. In light of this, the patient was transferred to the pediatric intensive care unit, where he received noninvasive ventilation and a total of three doses of 0.5 mg naloxone with great response. Urinary laboratory test results showed a tramadol level of 38 μg/mL (positive drug level, ≥ 50 ng/mL), as well as urinary concentrations of O-desmethyltramadol and N-desmethyltramadol at 24 μg/mL (positive drug level, ≥ 100 ng/mL) and 4.6 μg/mL (positive drug level, ≥ 100 ng/mL), respectively. CYP genotype showed three functional alleles corresponding to CYP2D6*2 × 2/CYP2D6*2, which is consistent with an UM [26].

In another case report, a 66-year-old man with renal insufficiency suffered from severe respiratory depression after receiving tramadol via patient-controlled analgesia. Complete recovery was achieved after administration of naloxone indicating that respiratory depression was related to opioid toxicity. Genotyping showed that the patient was an UM with gene duplication. Higher concentration of O-desmethyltramadol, as well as slower clearance of drugs due to renal insufficiency, contributed to tramadol toxicity [27]. In 2017, the US Food and Drug Administration updated their warnings regarding codeine and tramadol use in the pediatric population, making their use contraindicated in patients under the age of 12 years [28]. Recommendations of codeine use are summarized in Table 4.5.

CYP2D6 Gene Polymorphism and Therapy with Oxycodone or Hydrocodone

Oxycodone is a potent semisynthetic μ opioid agonist analgesic, which is metabolized by both CYP3A and CYP2D6. In vitro and in vivo studies have demonstrated that oxycodone metabolism proceeds through four

TABLE 4.5
Codeine Therapy Recommendation Based on CYP2D6 Polymorphism and Age

Subjects	Recommendation	Commonly Associated CYP2D6 Allele
Poor metabolizer	Avoid codeine use due to poor conversion of codeine into morphine.	CYP2D6*4/*4 CYP2D6*4/*5 CYP2D6*4/*6 CYP2D6*5/*5
Intermediate metabolizer	Codeine may be ineffective in pain control due to poor conversion of codeine into morphine	CYP2D6*4/*10 CYP2D6*5/*41 CYP2D6*3/*41
Extensive metabolizers	Use of codeine is appropriate	CYP2D6*1/*1 CYP2D6*1/*2 CYP2D6*1/*4 CYP2D6*1/*10 CYP2D6*1/*41 CYP2D6*2/*2 CYP2D6*2/*5
Ultrarapid metabolizer	Codeine use should be avoided due to toxic levels of morphine.	CYP2D6*1/ *1 × n CYP2D6*1/ *2 × n
Children under 12 years of age	FDA recommends not to use codeine or tramadol in children	All genotype

metabolic pathways: N-demethylation, O-demethylation, 6-ketoreduction, and glucuronide conjugation. CYP3A catalyzes the N-demethylation to the major (80%) circulating metabolite, noroxycodone, whereas CYP2D6 catalyzes the O-demethylation to oxymorphone, which accounts for 10% of oxycodone metabolites. The additional O-demethylation of noroxycodone results in the formation of noroxymorphone, which is mediated by CYP2D6. Oxymorphone is 14 times more potent than oxycodone. Affinities for μ-opioid receptor is in order of highest to lowest, as oxymorphone > morphine > noroxymorphone -> oxycodone > noroxycodone. In a study based on 10 healthy volunteers, the authors observed that oxymorphone Cmax was 62% and 75% lower in PMs compared with EMs and UMs. Noroxymorphone Cmax reduction was even more pronounced (90%). In UMs, oxymorphone and noroxymorphone concentrations increased, whereas noroxycodone exposure was reduced by approximately 50% [29]. Balyan et al. investigated oxycodone metabolism in 30 children who received oral oxycodone postoperatively. Plasma levels of oxycodone and oxymorphone and CYP2D6 genotype were analyzed. The authors observed that compared with PM or IM, significantly greater oxymorphone exposure was seen in EM who also showed better pain control [30].

Klimas et al. commented that oxycodone itself is responsible for 83.02% and 94.76% of the analgesic effect after oral and intravenous administration, respectively. Oxymorphone, which has a much higher affinity for the μ-receptor, only plays a minor role (15.77% and 4.52% after oral and intravenous administration, respectively). Although the *CYP2D6* genotype modulates oxymorphone concentrations, oxycodone remains the major contributor to the overall analgesic effect and there is a minimal analgesic effect from these metabolites [31].

Hydrocodone is a semisynthetic opioid, structurally related to codeine. Hydrocodone undergoes CYP-dependent oxidative metabolism to the O- and N-demethylated products, hydromorphone and norhydrocodone, respectively. The μ-opioid receptor binding affinity of hydromorphone is 30 times greater than that of hydrocodone. The O-demethylation of hydrocodone is predominantly catalyzed by CYP2D6 and to a lesser extent by an unknown low affinity CYP enzyme. Norhydrocodone formation is attributed to CYP3A4. Comparison of recalculated published clearance data for hydrocodone, with those predicted in the present work, indicate that about 40% of the clearance of hydrocodone is via non-CYP pathways [32].

Hydrocodone is approximately 12 times more potent at the μ-opioid receptor than codeine. Almost half of total hydrocodone clearance is determined by oxidative metabolism via CYP2D6 to hydromorphone and CYP3A4-mediated conversion into norhydrocodone. These metabolites undergo conjugation by UGTs before excretion in the urine. The evidence between genetic polymorphism of CYP2D6 and hydrocodone adverse effects is quite scant. A developmentally delayed girl aged 5 years 9 months (35 kg) of Somali descent, treated with valproic acid (250 mg, 2 times per day) since birth for seizures, presented to her family doctor with a cold that had persisted for several days. She was prescribed hydrocodone bitartrate (1 mg/mL) to be taken one teaspoonful orally three times per day for 5 days for cold symptoms. She was also prescribed clarithromycin for an ear infection. The next morning the child was given another dose of hydrocodone before school. At 9:45 a.m., she was picked up from school by her uncle because she was lethargic and dizzy. At home, she was given hydrocodone at 10:30 a.m. and 3:00 p.m. (total dose: 30 mg over 24 h). Twelve hours later, at 3:00 a.m., the child was snoring hard during sleep. The mother was unable to wake up the child to take medication, so she was left to sleep. At 6:58 a.m., the mother found the child unresponsive and called 911. Paramedics found the child stiff and in asystole. No cardiopulmonary resuscitation was performed; she was transferred to the hospital, where she was pronounced dead. During autopsy, her blood hydrocodone concentration was found to be 140 ng/mL, but hydromorphone concentration was below the detection limit of the assay. The concentration of valproic acid was 43 μg/mL. No other drugs or metabolites were detected. Genetic analysis revealed that the child had a reduced capability to metabolize hydrocodone via the CYP2D6 pathway (*CYP2D6*2A/*41*). Coadministration of clarithromycin (a potent cytochrome P450 3A4 inhibitor) for an ear infection and valproic acid for seizures since birth further prevented drug elimination from the body. This case highlights the interplay between pharmacogenetic factors, drug-drug interactions, and dose-related fatality in a child [33].

Methadone toxicity and pharmacogenetics issues

Methadone is metabolized by both CYP3A4 and CYP2B6 isoenzymes, but CYP2D6 has a minor role. Genetic polymorphism is the cause of high interindividual variability for a given dose, for example, in order to achieve a methadone plasma concentration of 250 ng/mL, dosage of racemic methadone mixture may vary from 55 to 921 mg per day for a 70 kg patient without receiving any comedication [34]. Kharasch et al. showed that CYP2B6 genotypes affected methadone plasma concentrations, clearance, and metabolism. The carriers of *CYP2B6*6* allele (PMs) particularly homozygous had higher methadone concentration and slower elimination, compared with the homozygous for wild-type gene (*CYP2B6*1/*1*). However, carriers of *CYP2B6*4* allele had lower methadone concentration and faster elimination even compared with individuals carrying wild-type gene. In general, *CYP2B6* variants had greater influence on metabolism of (S)-methadone than (R)-methadone when administered orally compared to intravenous administration. Methadone metabolism and clearance may be significantly lower in African Americans due to larger proportion of *CYP2B6*6* carriers [35].

Levran et al. reported that carriers of *CYP2B6*6* allele metabolized methadone slower and as a result required a relatively low methadone dosage (<100 mg/day) during treatment of heroin addiction. However, no association was observed between methadone dosage and polymorphism of *CYP3A4* or *CYP2D6* genes [36]. In a study of 40 postmortem cases where methadone was involved, the authors observed that *CYP2B6*6* allele was associated with highest postmortem methadone concentration (*CYP2B6*6/*6*, homozygous, postmortem methadone concentration: 1.38 mg/L) [37]. Madadi et al. reported two fatalities in infants related to methadone use. An exclusively breastfed 3-week-old male infant was born after 36 weeks of gestation. The mother was receiving 65 mg methadone per day during pregnancy and breastfeeding. The infant was unresponsive 3 h after nursing and was brought to the hospital where he later died. Forensic analysis of postmortem heart blood showed methadone level of 79 ng/mL. The infant was found to be *CYP2B6*6* haplotype, which has been associated with methadone-related mortality in adults due to impaired ability to metabolize methadone. In the second case, an 18-year-old male infant was born at 35 weeks of gestation to a mother who was prescribed methadone and was also abusing cocaine and smoking cigarette during pregnancy. The infant remained in hospital for 17 days and was discharged but received 0.1 mg dose of morphine for withdrawal symptoms. At home, he was sick after breastfeeding and was admitted to the hospital but later died. Both methadone and benzoylecgonine (metabolite of cocaine) were detected in a baby bottle containing breast milk. The neonate's postmortem heart blood showed methadone at a level of 26 ng/mL. Methadone concentration in the mixed blood was 33 ng/mL. Naloxone was also detected. The infant was also heterozygous for the three SNPs in ABCB1 associated with decreased P-glycoprotein activity. This efflux transporter is expressed in the luminal membrane of the blood-brain barrier and functional impairment in its activity has been shown to increase amount of methadone that reaches the brain causing severe toxicity [38]. In another study, the fatal blood methadone levels in three infant under 1 year of age ranged from 69 to 700 ng/mL as confirmed by LC-MS/MS [39].

Methadone is a substrate for P-glycoprotein encoded by *ABCB1* gene. It has been suggested that polymorphism of *ABCB1* gene may influence methadone kinetics. However, recently, Mouly et al. concluded that methadone dose was predicted by sociodemographic variables (bodyweight, fractioned dose, past cocaine dependence, and ethnicity), as well as clinical variables rather than genetic polymorphism [40]. However, other studies have shown some correlation between methadone dose and polymorphism of *ABCB1* gene, which is discussed later in the chapter.

Morphine and pharmacogenetic considerations

About 10%–30% of patients who receive morphine do not have adequate pain relief either because of inadequate analgesia or intolerable adverse effects. Genetic polymorphism of enzymes conjugating morphine may be responsible for such variations. Morphine is conjugated by Phase II enzyme UGT into morphine-3-glucuronide and morphine-6-glucuronide. The isoenzyme UGT2B7 is the major enzyme responsible for glucuronidation of morphine, while UGT1A1 plays a minor role. There are conflicting reports regarding association between *UGT2B7* polymorphism and metabolism of morphine. Basatami et al. analyzed blood specimens collected from 40 patients following abdominal hysterectomy, 24 h after initiation of analgesia through a patient-controlled analgesia pump. The specimens were genotyped and analyzed for morphine and its metabolites. In addition, the authors also genotyped approximately 200 autopsies found positive for morphine in routine forensic analysis. Patients homozygous for *UGT2B7 802C* needed significantly lower dose of morphine for pain relief. The same trend was observed for patients homozygous for *ABCB1 1236T* and *3435T*, as well as to *OPRM1 118A*. The dose of morphine in patients included in this study was significantly related to variation in *UGT2B7 T802C*. Age was significantly related to both dose and concentration of morphine in blood. Regression analysis showed that 30% of differences in variation in morphine dose could be explained by SNPs in these genes. The authors concluded that gene polymorphisms contribute significantly to the variation in morphine concentrations observed in individual patients. In addition, patients who were homozygous for

UGT2B7 802 C genotype needed significantly lower dose of morphine for pain relief [41]. However, in another study using healthy volunteers, the authors found no association between different genotypes of OCT1, ABCB1, and UGT2B7 and morphine pharmacokinetics and pharmacodynamics [42].

Hydromorphone and pharmacogenomics

Hydromorphone is conjugated by UGT2B7 enzyme. However, polymorphism of UGT2B7 gene has no significant effect on pharmacokinetics or effectiveness of pain management in Taiwanese subjects [43]. Xia et al. performed an exploratory study to investigate potential link between hydromorphone therapy and pharmacogenomics. Specimen collected from adult emergency department patients with acute pain deemed to require intravenous opioids received 1 mg of intravenous hydromorphone. Primary outcome was pain score (numeric rating scale, NRS) reduction between baseline and 30 min after medication administration. Secondary outcomes were pain relief, patient satisfaction with analgesia, desire for more analgesics, and side effects (nausea, vomiting, and pruritis). SNPs in OPRM1 gene (opioid receptor, A118G), ABCB1 gene (opioid transporter, C3435T), COMT gene (pain sensitivity, G1947A), and UGT2B7 gene (opioid metabolism, -G840A) were tested. The authors screened 1438 patients for this study with mean age of 39 years in the study population. Sixty-three percent were female. This exploratory study did not show a significant difference in pain NRS reduction among patients carrying different SNPs. Patient satisfaction with analgesia and nausea were statistically significantly associated with OPRM1 and UGT2B7 gene polymorphisms, respectively [44].

Pharmacogenomics issues with fentanyl

Fentanyl is extensively metabolized by N-dealkylation to norfentanyl that is mediated by CYP3A4 and CYP3A5 (CYP3A) enzymes. Less than 1% is metabolized by alkyl hydroxylation, N-dealkylation, or amide hydrolysis to despropionylfentanyl, hydroxynorfentanyl and hydroxynorfentanyl. Drug interaction with CYP3A4 inhibitors resulted in mixed responses. Severe and even lethal fentanyl intoxications have been reported for fentanyl when given in combination with strong CYP3A inhibitors. Coadministration with ritonavir, a very potent irreversible CYP3A inhibitor, resulted in almost twofold increase in exposure to fentanyl, whereas other potent reversible CYP3A inhibitors resulted in modest increases (26%–39%), or in case of itraconazole, no significant effect on fentanyl exposure was observed. Studies investigating effects of polymorphism of CYP3A gene on metabolism of fentanyl produced mixed results especially for CYP3A5*3 allele. This SNP results in reduced activity of the CYP3A5 enzyme and was linked with a twofold increase in systemic exposure to fentanyl. Fentanyl exerts its effect via binding with the μ-opioid receptor. It is 60–100 times more potent than morphine. Probably, the most important gene in relation to fentanyl pharmacodynamics is the OPRM1 gene. This gene encodes for the μ-opioid receptor. The A118 SNP is the most studied variant in the OPRM1 gene. The variant allele was associated with decreased effects of opioids [45].

Kuip et al. commented that CYP3A5*3 gene polymorphism may influence fentanyl pharmacokinetics as well, although further study is warranted. Several other factors have been studied but did not show significant and clinically relevant effects on fentanyl pharmacokinetics. Unfortunately, most of the published papers that studied factors influencing fentanyl pharmacokinetics describe healthy volunteers instead of cancer patients. Results from the studies in volunteers may not be simply extrapolated to cancer patients because of multiple confounding factors. To handle fentanyl treatment in a population of cancer patients, it is essential that physicians recognize factors that influence fentanyl pharmacokinetics, thereby preventing potential side effects and increasing its efficacy [46]. Effects of genetic polymorphism on metabolism and toxicity of hydrocodone, hydromorphone, oxycodone, tramadol, fentanyl, methadone, and morphine are summarized in Table 4.6.

P-GLYCOPROTEIN AND OPIOIDS

Opioid pharmacokinetics can demonstrate variability by alteration of distribution by transporters or metabolism. P-glycoprotein 1 (permeability glycoprotein) also known as MDR1 or ATP-binding cassette subfamily B member 1 (ABCB1) is an important protein of the cell membrane that pumps many foreign substances, including drugs out of cells. The multidrug resistance gene ABCB1 (adenosine triphosphate–binding cassette, subfamily B, member 1) encodes P-glycoprotein. More than 40 SNP's of this gene has been identified, which alters expression of P-glycoprotein. Some of these alterations potentially allow greater concentrations of opioids to enter in the brain following distribution into the blood-brain barrier [47].

A specific SNP of the ABCB1gene, labeled C3435T (3435 C is the wild-type and 3435 T is the variant allele), has been shown to improve the efficacy of opioids in experimental pain. This same SNP has also

TABLE 4.6
Effect of Genetic Polymorphism on Metabolism and Toxicity of Hydrocodone, Hydromorphone, Oxycodone, Tramadol, Fentanyl, Methadone, and Morphine.

Drug	Enzyme	Comments
Hydrocodone	CYP2D6	Currently, there is insufficient information to conclude whether poor metabolizers can be expected to experience decreased analgesia or whether ultrarapid metabolizers are more susceptible to toxicity.
Hydromorphone	UGT2B7	Polymorphism of *UGT2B7* gene has no significant effect on disposition and metabolism of hydromorphone in Taiwanese subjects. However, in one study, the authors concluded that the patient satisfaction with analgesia and nausea was statistically significantly associated with polymorphisms of *OPRM1* and *UGT2B7*, respectively.
Oxycodone	CYP2D6	Because of conflicting reports, it is difficult to conclude whether *CYP2D6* polymorphism is associated with analgesic effect or risk of toxicity from oxycodone use.
Tramadol	CYP2B6	Tramadol is likely to have reduced clinical efficacy in poor metabolizers. Ultrarapid metabolizers show higher peak level of active metabolite O-desmethyltramadol and are at higher risk of toxicity. Tramadol should not be used in children under 12 years of age.
Fentanyl	CYP3A5	*CYP3A5*3* allele has reduced CYP3A5 enzymatic activity and may cause elevated levels of fentanyl after standard dose, but there are conflicting results.
Methadone	CYP2B6	Poor metabolizers who are carrier of *CYP2B6*6* require lower dosage of methadone (<100 mg/day), but higher methadone postmortem levels have been observed in carriers of *CYP2B6*6* allele. There are conflicting reports regarding effect of *CYP3A4* polymorphism, as well as genetic variability of *ABCB1* gene on methadone metabolism and toxicity.
Morphine	UGT2B7	There are conflicting reports regarding association between *UGT2B7* gene polymorphism and metabolism of morphine. Subjects with *OPRM1 A118G* polymorphism may require higher morphine doses for pain management than those with the AA homozygotes

been shown to decrease postoperative pain in pediatrics and to decrease opioid requirements for postoperative pain and cancer pain. This is consistent with a study that found C3435T to be associated with opioid-induced respiratory depression and case reports that linked it to respiratory depression in pediatric patients following adenotonsillectomy. Thus, the C3435T SNP likely decreases P-glycoprotein activity (ineffective efflux pump), which reduces efflux of opioids away from target receptors, thus increasing concentration at opioid receptors. However, there are some studies that failed to show a relationship between *ABCB1* gene polymorphism and postoperative pediatric pain scores, analgesic effect of oxycodone on postoperative pain, or morphine use in patients following caesarean section [48].

Mamie et al. prospectively followed a cohort of 168 children after orthopedic or abdominal surgery, who were under morphine patient-controlled analgesia. The children and their parents were genotyped for six candidate gene polymorphisms (SNPs) implicated in nociception and opiate metabolism: C3435T for

ABCB1 gene, Val158Met for *COMT* (encodes catechol-O-methyltransferase), His40Tyr for *NTRK1* (encodes neurotrophic receptor tyrosine kinase 1 protein), 118G for *OPRM1* (encodes μ-opioid receptor), Arg236Gln for *POMC* (encoding proopiomelanocortin), and a haplotype of *CYP2D6*. Postoperative pain was assessed using the Faces Pain Scale. *ABCB1* gene 3435C to T polymorphism results in three different genotypes (CC, CT, and TT). The authors observed that pain peaks were more frequent in children with CC (wild-type) than in children with CT and TT, as well as were more frequent in children with *OPRM-GA* (transition of and adenine (A) nucleotide to guanine (G) at base 118) than those with *OPRM-AA*. The authors concluded that *ABCB1* and *OPRM 1* genotypes are associated with clinically meaningful pain variability, whereas *NTRK1* and *COMT* are linked to subclinical effects [49].

Candiotti et al. using 152 patients who were undergoing a nephrectomy reported that based on a mixed linear model, the *ABCB1* three genotypes showed a statistically significant effect on opioid. The TT genotype had significantly lower levels of cumulative opioid consumption compared with the CC genotype in first 24 h after surgery. The authors concluded that there is an association between the *ABCB1* polymorphism (C3435T) and interindividual variations in opioid consumption in the acute postoperative period after nephrectomy. The *ABCB1* gene polymorphism may serve as an important factor to guide acute pain therapy in postoperative patients [50].

Park et al. reported that in Korea, patients carrying the linked 3435T and 2677T alleles showed a significant difference in the level of respiratory suppression compared with other genotypes. The authors concluded that analysis of *ABCB1* polymorphisms may have clinical relevance to prevent respiratory suppression by intravenous fentanyl or to anticipate its clinical effects [51]. However, Saiz-Rodríguez et al. based on a study of 35 healthy volunteers (19 men and 16 women) receiving a single 300 μg oral dose of fentanyl concluded that *CYP3A5*3*, *ABCB1* C3435T, and *ABCB1* G2677T/A were not associated with fentanyl's pharmacokinetics, pharmacodynamics, and safety profile [52].

Therefore, functional studies of the C3435T polymorphism demonstrate somewhat mixed outcomes and might be more predictive of adverse drug reactions. In surgical patients receiving morphine, there was no significant association between the C3435T polymorphism and postoperative morphine dose requirements; however, there was an association with the requirement for nausea/vomiting therapy [53]. Additionally, haplotype analysis for *ABCB1* gene may be more predictive for dose requirement. In therapy of opioid dependence with methadone, patients carrying two copies of the wild-type haplotype required higher methadone doses when compared with one or both variants [54]. Outcome differences in these studies may be due to additional population variation or sample size; however, it appears that haplotype or multiple gene analysis may provide an interpretative advantage for prediction of analgesia and adverse effects related to *ABCB1* variants. In one study, 99 patients undergoing abdominal surgery with colorectal anastomosis because of colorectal carcinoma were genotyped for C34535T. Then patients were divided into three groups according to their genotype: CC-wild type homozygous, CT-mutant heterozygous, and TT-mutant homozygous. Intravenous fentanyl patient-controlled analgesia was provided postoperatively for pain control in the first 24 h after surgery. Opioid consumption, pain scores, and the adverse side effects were evaluated. The authors observed that the patients in the CC genotype group consumed significantly more fentanyl

(375.0 μg ± 43.1) than the patients in the TT group (295.0 μg ± 49.1) and the CT (356.4 μg ± 41.8) group in the treatment of postoperative pain. The patients in the TT group had lower pain scores at 6, 12, 18, and 24 h postoperatively. There were no significant differences in the side effects among the three groups regarding the vomiting and the sedation score. The patients in the TT group had more frequently complained about nausea. The authors concluded that the C3435T SNPs of the *ABCB1* gene are associated with differences in the opioid sensitivity. The *ABCB1* polymorphism may serve as an important genetic predictor to guide the acute pain therapy in postoperative patients [55].

POLYMORPHISMS OF GENES CODING OPIOID RECEPTORS AND OPIOID THERAPY

Opioid receptors are a group of G protein–coupled receptors with opioids as ligand. The opioid receptors play important roles in analgesic effects of opioids. There are three classes of opioid receptors: μ, δ, and κ (mu, delta, and kappa), and these opioid receptors are widely expressed in the brain. There are three distinct families of opioid peptides: endorphins, encephalin, and dynorphins. Encephalin binds to δ receptor, dynorphins bind to kappa receptor, while endorphins bind to both μ and δ receptors. More recently, two additional short peptides endomorphin-1 and endomorphin-2 that display high affinity for μ receptors have been characterized. The opioid receptors are encoded by three specific genes: *OPRM1* at 6q24-q25 encodes the μ receptor, *OPRD1* at 1p36.1-p34.3 encodes the delta receptor, and *OPRK1* at 8q11.2 encodes the kappa receptor.

The μ1 receptor gene (*OPRM1*) that encodes the μ receptor has a functionally significant and common variant termed A118G (rs1799971). This SNP in exon 1 of the gene results in transition of and adenine (A) nucleotide to guanine (G) at base 118. The A118G transition in the DNA sequence causes the amino acid exchange at residue 40 of the μ-opioid receptor protein from the normal asparagine (Asn) to abnormal aspartic acid (Asp) residue (Asn40Asp). The Asp40 isoform does not carry N-glycosylation site in the extracellular region of the receptor that reduces expression of the isoform at the cell surface and also decreases binding potential of the receptor to opioids. As a result, carriers of this allele require higher morphine dose. A118G has a minor allele frequency of 38% in Asians, 16% in Europeans, and 3% in African Americans. Overall, 64% of the population has normally functioning *OPRMI* gene encoding μ-opioid receptor [56].

Yu et al. performed a metaanalysis of data of the 467 screened studies, 12 studies 2118 participants were eligible to be included in the analysis. The metaanalysis results indicated that G allele carriers (AG + GG) of the OPRM1 A118G polymorphism required higher opioid doses for pain management than those with the AA homozygotes (wild-type gene). In subgroup analysis, the authors did not find statistically significant correlation between OPRM1 A118G polymorphism and opioid pain relief among Caucasian patients, as well as among morphine users, except for Asian patients. The authors based on their metaanalysis concluded that G allele (AG + GG) carriers of OPRM1 A118G polymorphism required more opioid analgesia in cancer pain management [57].

Boswell et al. investigated the role of hydromorphone and OPRM1 in postoperative pain relief with hydrocodone by enrolling 158 women scheduled for Cesarean section in the study. The patients had bupivacaine spinal anesthesia for surgery and received intrathecal morphine with the spinal anesthetic or parenteral morphine for the first 24 h after surgery. Thereafter, patients received hydrocodone/acetaminophen for postoperative pain control. On postoperative day 3, venous blood samples were obtained for OPRM1 A118G genotyping and serum opioid concentrations. In the study population, 131 (82.9%) of the women enrolled in the study were homozygous for the 118A allele of OPRM1 (AA) and 27 (17.1%) carried the G allele (AG/GG). By regression analysis, pain relief was significantly associated with total hydrocodone dose in the AA group but not in the AG/GG group. In contrast, there was no association between pain relief and serum hydrocodone concentration in either group. However, pain relief was significantly associated with serum hydromorphone concentration (active metabolite of hydrocodone) in the AA group, but not in the AG/GG group. In addition, side effects were significantly higher in the AG/GG group than in the AA group, regardless of adjustment for body mass index, pain level, or total dose of hydrocodone. This study found a correlation between pain relief and total hydrocodone dose in patients homozygous for the 118A allele (AA) of the OPRM1 gene, but not in patients with the 118G allele (AG/GG). However, pain relief in 118A patients did not correlate with serum hydrocodone concentrations, but rather with serum hydromorphone levels, the active metabolite of hydrocodone. This suggests that pain relief with hydrocodone may be due primarily to hydromorphone. Although pain relief did not correlate with opioid dose in AG/GG patients, they had a higher incidence of opioid side effects. The correlations identified in this study may reflect the fact that

serum opioid concentrations were measured directly, avoiding the inherent imprecision associated with relying solely on total opioid consumption as a determinant of opioid effectiveness. Thus, measurement of serum opioid concentrations is recommended when assessing the role of OPRM1 variants in pain relief. This study supports pharmacogenetic analysis of OPRM1 in conjunction with serum opioid concentrations when evaluating patient responses to opioid therapy [58].

Most clinically used opioids are μ-opioid receptor agonists. Therefore, genetic variation of the OPRM1 gene that encodes the μ-opioid receptor is of great interest for understanding pain management. Cajanus et al. studied the association between the $118A > G$ polymorphism and oxycodone analgesia and pain sensitivity in 1000 women undergoing breast cancer surgery. The $118A > G$ variant was associated significantly with the amount of oxycodone requested for adequate analgesia. Overall, oxycodone consumption was highest in individuals having the GG genotype, lowest for those with the AA genotype, and moderate for those having the AG genotype. The authors concluded that the OPRM1 $118A > G$ polymorphism was associated with the amount of oxycodone required in the immediate postoperative period. Although a significant factor for determining oxycodone requirement, the $118A > G$ polymorphism alone explained less than 1% of the variance [59]. In contrast, Zwisler et al. observed no difference between OPRM1, the wild type, and the variant allele in the percentages of nonresponders (118AA = 16.4% vs. 118AG/ 118GG = 17.0%, $P = 1.0$) or in the pain ratings. For ABCB1, no difference was found between the wild-type and the variant alleles for SNP tested as percentages of nonresponders (3435CC = 17.5% vs. 3435CT/ 3435TT = 15.8%, $P = 0.85$; 2677GG = 17.8% vs. 2677GT/2677TT = 15.8%, $P = 0.74$) or pain ratings. The authors concluded that there was no association between the tested SNPs in OPRM1 and ABCB1 and changes in the analgesic effect of oxycodone [60].

OPRM1 Gene Polymorphism and Addiction
Haerian and Haerian based on metaanalysis of 13 studies (n = 9385), comprising 4601 opioid dependents and 4784 controls, observed significant association between OPRM1 A118G (rs1799971) polymorphism and susceptibility to opioid dependence in overall studies under a codominant model, as well as susceptibility to opioid dependence or heroin dependence in Asians under an autosomal dominant model. The nonsynonymous OPRM1 rs1799971 might be a risk factor for addiction to opioids or heroin in an Asian

population [61]. Chen et al. concluded that the *OPRM1* A118G polymorphism may contribute to the susceptibility of alcohol dependence in Asians but not in Caucasians [62].

Türkan et al. investigated the prevalence of the *OPRM1* A118G polymorphism (rs1799971) in Turkish population and its association with opioid and other substance addiction. In addition, the authors also examined the association of rs1799971 in addicted patients who were also diagnosed with psychiatric disorders. The study included 103 patients addicted to opioids, cocaine, ecstasy, alcohol, lysergic acid diethylamide, cannabis, and sedative/hypnotic substances and 83 healthy volunteers with similar demographic features as controls. The rs1799971 polymorphisms were identified with the polymerase chain reaction restriction fragment length polymorphism method. The genotype frequencies were significantly higher in the addicted patients than controls (32.0% vs. 16.9%, respectively; $P = .027$). The prevalence of the G allele was 16.1% in the addicted group and 8.4% in the control group ($P = .031$). The authors concluded that there was an association between the rs1799971(G) allele frequency and opioid and other substance addiction, but no association was found between this polymorphism and psychiatric disorders [63].

DOPAMINERGIC SYSTEM AND ADDICTION

The monoaminergic neurotransmitter systems constitute the primary reward pathway in the human brain. Monoamines can be further divided into catecholamines (dopamine epinephrine and norepinephrine) and serotonin. Dopamine is synthesized from tyrosine, which plays an important role in regulating emotional and motivational behavior through the mesolimbic dopaminergic pathway which also includes rewards after consuming food and also reward related to administration of drugs of abuse. Dopamine neurons are located in the VTA of the brain and projecting into the NAc. After receiving the appropriate signal, dopamine is released by presynaptic neuron into the synapse where dopamine binds with dopamine receptor. The unbound dopamine is then taken up by d DATs for future utilization or degradation.

The mesolimbic dopaminergic pathway is the major reward pathway of the brain. As a result, dysfunctions in the dopaminergic systems are involved in several pathological conditions, including Parkinson's disease, Tourette's syndrome, drug addiction, and hyperactivity disorders. Dopamine binds to two families of G protein–coupled receptors. The D1-type receptors are D1 (DRD1) and D5 receptors (DRD2), while D2-type receptors are DRD2, DRD3, and DRD4. The diverse physiological functions of dopamine receptors encoded by the genes *DRD1*, *DRD2*, *DRD3*, *DRD4*, and *DRD5* and polymorphisms of these genes can results in change in the expression levels and may be associated with susceptibility to alcohol and or drug abuse. The cell bodies of dopaminergic neurons are located in the hypothalamus and the brain stem region (substantia nigra, pars compacta, and VTA). Dopamine is involved in regulation of movement, attention, motivation, and reward from various activities, including intake of food and abuse of alcohol and or drugs.

DAT belongs to the family of sodium/chloride-dependent transporter proteins and is responsible for the reuptake of extracellular synaptic dopamine into presynaptic neurons, thus terminating the effect of dopamine. Therefore, dopaminergic reward circuit may function differently when DAT expression level alters. The *SLC6A3* gene present on chromosome five (5q15.3) encodes DAT responsible for sequestering dopamine back into presynaptic neurons. Dopamine-β-hydroxylase (DβH), encoded by *DβH* gene, catalyzes the conversion of dopamine to norepinephrine, and several polymorphisms in the *DβH* gene have been the focus of addiction research for many years. The enzyme COMT is involved in the metabolism of catecholamines, including dopamine. Variation in COMT activity may influence dopamine levels in the prefrontal cortex (PFC), which may lead to alteration in reward process, as well as susceptibility to drug addiction.

All abused drugs exert their initial reinforcing effects by triggering supraphysiological surges of dopamine in the NAc either by direct or indirect mechanism. Repeated drug administration results in neuroplastic changes in the brain [64]. While drugs of abuse increase dopamine levels in the NAc, most drugs that are not habit forming do not cause dopamine overflow. Psychostimulants increase dopamine levels primarily by altering dopamine clearance from extracellular space, whereas opiates indirectly increase dopamine transmission by suppressing inhibitory output onto dopamine neurons [65]. In general, stimulants such as amphetamine, methamphetamine, and cocaine have their most pronounced effects on monoamine neurotransmitters such as dopamine, norepinephrine, epinephrine, and serotonin. Acute administration of amphetamine-like drugs directly stimulates release of dopamine from neurons. However, repeated administration of these drugs results in tolerance, which is related to depletion of dopamine in the presynaptic terminals. As a result, withdrawal symptoms develop when the drug is discontinued after prolonged use.

Dopaminergic System and Opioid Addiction

The dopaminergic system is known to mediate drug reward and reinforcement. Therefore, polymorphisms of genes encoding dopamine receptors may be associated with drug addiction. Dopamine D1 receptor (DRD1) modulates opioid reinforcement, reward, and opioid-induced neuroadaptation. Zhu et al. hypothesized that *DRD1* gene polymorphism may be associated with susceptibility to opioid dependence, the efficiency of transition to opioid, and opioid-induced pleasure response. The authors analyzed potential association between seven *DRD1* polymorphisms with the following traits: duration of transition from the first use to dependence, subjective pleasure responses to opioid on first use and postdependence use, and risk of opioid dependence in 425 Chinese with opioid dependance and 514 healthy controls. Most addicts (64.0%) reported noncomfortable response upon first opioid use, while after dependence, most addicts (53.0%) felt strong opioid-induced pleasure. Survival analysis revealed a correlation of prolonged duration of transition from the first use to dependence with the minor allele—carrying genotypes of *DRD1* rs4532 and rs686. The authors concluded that *DRD1* rs686 minor allele decreases the opioid dependance in China [66].

Liu et al. also examined whether dopamine receptor D1 (DRD1) is associated with heroin dependence and the impulsive behavior in patients with heroin dependence. The participants included 367 patients with heroin dependence and 372 healthy controls from a Chinese Han population. The authors examined the potential association between heroin dependence and eight SNPs (rs686, rs4867798, rs1799914, rs4532, rs5326, rs265981, rs10078714, rs10078866) of *DRD1* and the associations between single SNP, haplotypes, and impulsive behavior. Compared with the healthy controls, heroin-dependent patients showed a significantly lower frequency of GG homozygotes of rs5326, significantly lower frequency of the G allele of rs5326, and higher frequency of the rs265981 G allele. The authors concluded that the two SNPs, rs5326 and rs265981, were not associated with the impulsive behavior in patients with heroin dependence [67].

The *DRD2* gene shows primarily three kinds of polymorphism, including −141 c ins/del, Taq 1A, and Taq1B alleles. Taq1A is also known as TaqIA (rs1800497). The *DRD2* Taq1A (A1 allele causing lower DRD2 receptor density in striatum) polymorphism is located 10 kilobase pairs downstream from the coding region of the *DRD2* gene at chromosome 11q23. In fact, the *DRD2*-associated polymorphism has been more precisely located within the coding region of

ANKK1 (ankyrin repeat and kinase domain containing 1) gene, which is a neighboring gene that may confer a change in amino acid sequence. One functional SNP in dopamine receptor D2 (DRD2), rs1076560, is involved in regulating splicing of the gene and alters the ratio of DRD2 isoforms located pre- and postsynaptically. The rs1076560 has been previously associated with cocaine abuse. Clarke et al. analyzed the role of rs1076560 in opioid dependence by genotyping European American (n = 1041) and African American (n = 284) opioid addicts and reported that DRD2 SNP rs1076560 is associated with opioid addiction [68].

Several polymorphism of DRD3 receptor gene have been reported to be associated with susceptibility with alcoholism. Although *DRD3* rs6280 (Ser9Gly; substitution of a glycine for serine residue in the extracellular receptor N-terminal domain also known as Bal I restriction due to a restriction site produced by the variant) polymorphism which makes the receptor more sensitive to dopamine is associated with various psychiatric disorders. In addition, Bal I restriction is associated with greater risk of cocaine abuse, as well as nicotine abuse [69]. Duaux et al. proposed that homozygosity for the Bal I polymorphism of *DRD3* gene is associated with predisposition to opioid dependance [70].

The DRD4 receptors, G-coupled receptors encoded by the *DRD4* gene, have been linked to several psychiatric disorders (schizophrenia, bipolar disorder), neurological disorders (Parkinson's disease), and addictive behaviors (including novelty seeking). DRD4 has a "variable number of tandem repeat" (VNTR) polymorphism on exon 3 that may affect gene expression by binding nuclear factors. The *DRD4* VNTR can vary from 2 to 10 with 48-bp repeats where alleles with 6 or less than 6 repeats are termed as "short alleles" and alleles with 7 or more than 7 repeats are termed as "long alleles." Compared with short alleles, long alleles may reduce *DRD4* gene expression. The most studied polymorphism of *DRD4* gene is a 48 base pair VNTR in the third axon. In general, *DRD4* VTNR long allele with 7 or more repeats is associated with novelty seeking and drinking behavior. A seven repeat polymorphism at exon 3 was observed more frequently in methamphetamine abusers than controls [71].

Shao et al. tested the hypothesis that heroin addicts carrying D4 dopamine receptor gene (DRD4) VNTR long-type allele would have higher craving after exposure to a heroin-related cue. Craving was induced by a series of exposure to neutral and heroin-related cue and was assessed in a cohort of Chinese heroin abusers (n = 420) recruited from the Voluntary Drug Dependence Treatment Center at Shanghai. The authors reported that significantly stronger cue-elicited heroin craving was found in individuals carrying *DRD4 VNTR* long-type allele than the noncarriers. The author concluded that *DRD4 VNTR* polymorphism contributes to cue-elicited craving in heroin dependence, indicating *DRD4 VNTR* represents one of the potential genetic risk factors for cue-induced craving [72].

It is generally accepted that dopaminergic system plays an important role in the rewarding effects of drugs of abuse. Heroin and other opioid receptor agonists increase dopamine levels indirectly by inhibiting GABAergic neurons. Randesi et al. studied whether dopaminergic system gene variants are associated with opioid dependence by recruiting 153 controls, 163 opioid exposed, but not dependent, and 281 opioid-dependent subjects. The authors examined genotypes of 90 variants in 13 genes and concluded that dopamine-β-hydroxylase variants rs2073837 and rs1611131 if present protect individuals from opioid abuse. In contrast, tyrosine hydroxylase variant rs2070762 increase the risk of opioid addiction [73].

Growing evidence indicates conflicting results about the DRD2/kinase domain containing 1 gene (*ANKK1*) *TaqIA* (*Taq1A*) SNP (rs1800497) and increased risk of drug dependence, including opioid dependance. Deng et al. performed metaanalysis of a total of 25 available studies and observed that the *DRD2/ANKK1 TaqIA* polymorphism was significantly associated with increased risk of opioid dependence under homozygote, dominant, and recessive genetic model, respectively. The authors concluded that their current metaanalysis suggested that *DRD2/ANKK1 TaqIA* polymorphism might be associated with opioid dependence risk, but not associated with stimulants or marijuana dependence [74]. A haplotype block of 25.8 kilobases (kb) was defined by eight SNPs extending from SNP3 (TaqIB) at the 5′ end to SNP10 site (TaqIA) located 10 kb distal to the 3′ end of the gene. Xu et al. concluded that within this block, specific haplotype cluster A (carrying TaqIB1 allele) was associated with a high risk of heroin dependence in Chinese patients [75].

COMT degrades catecholamines (any of a group of sympathomimetic amines, including dopamine, epinephrine, and norepinephrine). A variant of *COMT* gene (val158met), where a valine is substituted by methionine at codon 158 is produced by SNP G772A has reduced activity which increases dopamine levels. This increased level of circulating dopamine suppresses endogenous opioid production, which in turn upregulates opioid receptors. This increase in opioid binding sites in response to this variant has been seen in postmortem brain cells, and val158met has been shown to decrease regional/endogenous μ-opioid system response and cause higher pain ratings to experimental pain. An early study demonstrated val158met heterozygosity as being linked to decreased opioid requirements for cancer pain. These findings were later reproduced for postoperative pain. However, other studies failed to show significance of the val158met variant as produced by the SNP G772A and instead showed significance in certain COMT haplotypes, which are a set of polymorphisms that have a tendency toward being inherited together. These haplotypes were found to modulate COMT activity [48].

Zhang et al. recruited 115 Chinese patients who received fentanyl analgesia during first 48 h after surgery. The authors selected Han Chinese patients undergoing radical gastrectomy to evaluate whether the genetic polymorphisms of rs6269, rs4633, rs4818, and rs4680 and their haplotypes affect postoperative fentanyl requirements for analgesia. Among SNPs, rs6269 is located at the promoter region 9 and the other 3 SNPs (rs4633, rs4818, and rs4680) are located at the coding region of the *COMT* gene. The variations of rs6269, rs4633, and rs4818 are synonymous mutations, whereas the variation of rs4680 is a missense mutation and leads to the substitution of valine to methionine (val158met). The rs6269 polymorphism of the *COMT* gene can form 3 genotypes: AA, AG, and GG. Among 115 patients in the study population, 47 patients were carriers of AA genotype, 48 patients with the AG genotype, and 20 patients with the GG genotype. Thus, the frequency of rs6269 G allele was 38.3% in this study. Similar to rs6269 polymorphism, rs4633 polymorphism also develops 3 genotypes: CC, CT, and TT. Of the 115 patients, 50 patients were carriers of the CC genotype, 51 patients with the CT genotype, and 14 patients with the TT genotype, with the frequency of the T allele being 34.3% in rs4633 polymorphism. In this study, rs4818 polymorphism generated 3 genotypes; there were 20 carriers of the GG genotype by endonuclease digestion, 49 patients with the CC genotype, 46 patients with the CG genotype, and no T allele mutant by DNA sequencing; the frequency of the G allele was 37.4%. The rs4680 polymorphism forms GG, GA, and AA genotypes. There were 55 carriers of the GG genotype, 47 patients with the GA genotype, and 13 patients with the AA genotype; the frequency of the A allele was 31.7% in rs4680 polymorphism. The allele frequency of rs6269, rs4633, rs4818, and rs4680 was found to be in Hardy–Weinberg equilibrium (all $P > .295$). There were no significant differences of fentanyl doses among different SNPs of *COMT* rs6269, rs4633, rs4818, and rs4680 at 24 and 48 h after surgery. However, *COMT* gene haplotypes combined by *COMT* rs6269, rs4633, rs4818, and rs4680 significantly affected fentanyl doses at 24 and 48 h after surgery. Among these 3 haplotypes (GCGG, ATCA, and ACCG), patients with haplotype patients with haplotype ACCG consumed more fentanyl than GCGG and ATCA haplotypes during the first 24 and 48 h after surgery. No significant differences were found in the incidence of nausea, vomiting, and dizziness among the four SNPs of *COMT* gene and their haplotypes. The authors concluded that *COMT* gene haplotype constructed by rs6269, rs4633, rs4818, and rs4680 contributes to the individual variation of postoperative analgesia with fentanyl. Patients carrying the *COMT* gene haplotype ACCG consumed the highest amount of fentanyl during the first 24 and 48 h postoperatively [76]. Links between dopaminergic system and opioid abuse are listed in Table 4.7.

TABLE 4.7
Association Between Polymorphism of Dopaminergic System Gene and Opioid Abuse.

Gene	Polymorphism	Effect on Opioid Addiction
Dopamine receptor 1 (*DRD1*)	*DRD1* rs686 minor allele	Reduced risk of opioid dependance in Chinese population.
Dopamine receptor 2 (*DRD2*)	(*DRD2*) SNP rs1076560 *DRD2/ANKK1* TaqIA	Increases risk of opioid addiction. Increased risk of opioid addiction.
Dopamine receptor 3 (*DRD3*)	*DRD3 rs6280* (C allele: Ser9Gly)	Increased risk of opioid dependance.
Dopamine receptor 4 (*DRD4*)	Variable number tandem repeat (VNTR) long-type allele of DRD4 gene	May increase craving for heroin.
Dopamine-β-hydroxylase (*DBH*)	rs2073837 rs1611131	Both SNPs protect from opioid addiction.
Catechol-O-methyl transferase (*COMT*)	*COMT* gene haplotype constructed by rs6269, rs4633, rs4818, and rs4680	May affect required fentanyl dosage.
Tyrosine hydroxylase (*TH*)	rs2070762	Increased risk of opioid addiction.

OTHER NEUROTRANSMITTER PATHWAYS AND OPIOID ADDICTION

Randesi et al. investigated potential association between serotonergic and noradrenergic gene variants with heroin addiction. The authors analyzed a total of 126 variants in 19 genes in subjects with Dutch European ancestry from the Netherlands. These subjects included 281 opioid-dependent volunteers in methadone maintenance or heroin-assisted treatment, 163 opioid-exposed but not opioid-dependent volunteers who have been using illicit opioids but never became opioid-dependent, and 153 healthy controls. The authors reported significant association from the combined effect of three *SLC18A2* (solute carrier family 18 member A2) gene polymorphisms (rs363332, rs363334, and rs363338) with heroin dependence. SLC18A2 is a protein encoded *SLC18A2* gene, and this protein is also known as the vesicular monoamine transporter 2, which is an integral membrane protein that transports monoamines, particularly neurotransmitters such as dopamine, norepinephrine, serotonin, and histamine from cellular cytosol into synaptic vesicles. The authors in this paper concluded that further studies are warranted to confirm and elucidate the role of these variants of *SLC18A2* gene in the vulnerability to opioid addiction [77].

Glutamic acid decarboxylase (GAD) is the rate-limiting enzyme in the conversion of glutamate to GABA. Two isoforms of GAD have been identified; GAD1 and GAD2. GAD1 is associated with cytosolic GABA production and is responsible for maintaining basal GABA level, while GAD2 is mostly involved in synaptosomal GABA release and can be rapidly activated on demand for GABA. However, GAD1 is the primary rate-limiting enzyme regulating GABA level under normal circumstances. The *GAD1* gene encodes the 67-kDa GAD. Wu et al. using 370 heroin-dependent subjects and 389 healthy controls observed that frequencies of the rs1978340 (promoter region), rs3791878 (promoter region), and rs11542313 (exon 3) polymorphisms in heroin addicts were significantly different from those in healthy controls. The authors concluded that in their case-control association study, the T alleles of *GAD1* rs1978340 and rs11542313 were strongly associated with decreased risk of heroin dependence, while the C allele of GAD1 rs3791878 was associated with increased risk of heroin dependence [78].

GENETIC ADDICTION RISK SCORE

Over 100 million Americans with addictive personalities have a genetic predisposition for addiction, which is called the Reward Deficiency Syndrome (RDS). These individuals have lower utilization of neurotransmitter chemicals associated with the brain reward system, and as a result, these people have "neurotransmitter deficits" compared with normal levels. This puts them at a disadvantage and makes them prone to accidents, aberrant cravings, and drug-seeking behaviors. Most drug-seeking behaviors originate in the dopaminergic centers of the mesolimbic region of the brain because these pathways are responsible for the feelings of pleasure and a sense of well-being. Any deficits or decrease in the dopaminergic system will lead to a loss of pleasure and eventually lead to drug-seeking or high-risk behaviors [79]. A genetically dependent decline in the number of receptors for neurotransmitters will lessen or attenuate the neurological reward/pleasure signal to the affected target organs, known as "dopamine resistance." The dopamine resistance creates a lower sense of well-being. DNA gene testing can identify these individuals who carry the affected Reward Deficiency Genes [80]. Genetic Addiction Risk Score (GARS) has been proposed as a measure of addiction vulnerability of an individual [79]. It has been awarded a US patent.

Geneus Genomic Testing Center, in conjunction with Dominion Diagnostics and Colorado University, has developed a GARS based on 11polymorphisms and 10 genes that may be associated with RDS. Interestingly, GARS score as a predictor of high risk for RDS behaviors has been associated with the Addiction Severity Index, which is a clinical predictive not diagnostic test. However, GARD score cannot display false positives because the GARS test measures an entire panel of gene polymorphisms to predict drug and alcohol severity as a cluster [81]. If a patient has four GARS risk alleles, then it may indicate susceptibility to drug addiction. In addition, if any combination of seven GARS risk alleles is presents, the test is predictive of alcohol severity. These 10 genes are listed in Table 4.8. Although some of these genetic polymorphism associated with substance abused focusing on heroin and opioid abuse have been discussed earlier in the chapter, these 10 genes are also summarized below.

Dopamine D1 receptor is encoded by *DRD1* gene. One polymorphism of this gene (*DRD1*-48 A > G) is associated with severity of alcohol-related problems The G allele is normal, whereas A allele is associated with alcohol-related problem, as well as novelty seeking behavior. The A-48G polymorphism, which is located in the 5'-untranslated region (UTR) of the *DRD1* gene, is one of the common polymorphism sites of the *DRD1* gene. The genetic polymorphism distribution usually varies in different ethnic groups. Asians have a

higher prevalence of A allele than that of Caucasians. Japanese subjects have even higher A allele prevalence than others. However, Liu et al. found no association found between *DRD1* A-48G polymorphism and methamphetamine abuse or methamphetamine-induced psychosis [82]. The A1 allele of the *DRD2/ANKK1 Taq1A* polymorphism (rs1800497) is associated with reduced striatal D2/3 receptor binding in healthy individuals, as well as depression and addiction [83]. This genetic polymorphism is included in GARS score. Association between polymorphism of dopamine receptor D3 gene (rs6280:C allele) and increased risk of drug addiction has been discussed in detail earlier.

The *DRD4* gene has a 48 bp VNTR polymorphism located in the third exon, and the 48 bp polymorphism can be repeated 2 to 11 times. Research suggests that at the molecular level, the 7R (7-repeat) allele may be associated with less efficient/unstable transcription and translation, less efficient folding and greater conformational rearrangement of receptor proteins, and weaker intracellular dopamine signaling. Because D4 receptors are densely distributed in areas associated with emotion, reward response, and motivation, including the hippocampus and the amygdala, carriers of 7R allele are more prone to alcohol and drug craving. Bobadilla et al. examined associations between the *DRD4* gene 48bp VNTR polymorphism and comorbidity between marijuana use frequency and depression in a diverse, nonclinical adolescent sample (n = 1882; ages 14–18) from the National Longitudinal Study of Adolescent Health. Multinomial regression analyses indicated that the odds of being comorbid for depressive symptoms and marijuana use are approximately 2.5 higher in youth who were carriers of 7R allele compared with youths not carrying this allele [84].

The gene for the human DAT1 displays several polymorphisms, including a 40-bp VNTRs ranging from 3 to 16 copies in the 3'-UTR of the gene. The 10-repeat (10R; A10 allele) and nine-repeat (9R; A9 allele) are more commonly observed than other polymorphisms. The 9R allele is associated with fast uptake of dopamine and may cause hypodopaminergic state, thus increasing risk of alcohol and drug abuse. Vasconcelos et al. based on a case-control study that included 227 males of northeastern Brazil (113 alcoholics and 114 controls) concluded that A9 allele and A9/A9 genotype of the *SLC6A3* 40-bp VNTR are involved in the vulnerability to alcohol dependance in the population studied [85]. The A9 allele of the DAT gene is also associated with delirium tremens and alcohol-withdrawal seizure [86].

Serotonin (5-hydroxytryptamine; 5-HT) is a monoamine neurotransmitter with multiple sites of action,

TABLE 4.8
Ten Genes Used for Determining Genetic Addiction Risk Score.

Name of the Gene	Risk Allele	Normal Allele	Comments
Dopamine D1 receptor (*DRD1*) gene	*DRD1−48A > G* (A allele)	G allele	*DRD1−48A > G* is associated with severity of alcohol-related problems. In addition, A/A genotype is also associated with novelty seeking behavior.
Dopamine D2 receptor (*ANKK1/DRD2*) gene	*TaqIA1*	A2 allele	*TaqIA1* allele is associated with low dopamine D2 receptor density and may cause hypodopaminergic functioning in dopamine-based reward pathway, thus increasing risk of alcohol/drug abuse.
Dopamine D3 receptor (*DRD3*) gene	*DRD3 rs6280* (C allele: Ser9Gly)	T allele	C allele is associated with increased risk of alcohol dependence and higher risk of cocaine, opioid, and nicotine abuse.
Dopamine D4 receptor (*DRD4*) gene	*DRD4- 7-repeat allele* (7R)	4 R allele	7R allele is associated with novelty seeking behavior.
Dopamine D transporter (*DAT1; SLC6A3*) gene	*DAT1-9-repeat allele* (9R allele).	The 40-bp VNTR element is repeated between 3 and 13 times in most people, but both 9 and 10 R allele are problematic	9R allele associated with fast uptake of dopamine may cause hypodopaminergic state. This allele may also increase risk of attention deficit hyperactive behavior (ADHD). The 10R allele is associated with slow uptake of dopamine and may cause hyperdopaminergic state.
Serotonin transporter (*5-HTTLPR*) gene	Short allele (S-allele)	Long allele (L-allele) is not counted	Increased risk of alcohol dependence in carrier of S-allele. The S-allele downregulates serotonin (of 5-HT1) and carriers of this allele have an increased analgesic response to opioids.
μ-Opioid receptor (*OPRM1*) gene	A118G (G allele)	A allele	G-allele increases risk of heroin abuse.
GABA receptor β3 subunit gene (*GABRB3*)	181-Bp	Wild type	The 181 SNP reduces sensitivity of GABA receptor and increases possibility of alcohol and drug abuse.
Gene encoding monoamine oxidase A enzyme	*MAO-uVNTR -4 repeat (4R) allele*	Wild type	4R allele is associated with fast catabolism of mitochondrial dopamine.
Gene encoding catechol amine methyl transferase (*COMT*) encoding gene	*COMT-Val58Met* (G allele)	A allele (met)	Val substitution (G allele) results in fast degradation of synaptic dopamine causing reward deficiency syndrome.

which can affect mood, sensory processing, cognition, and sleep. Serotonergic neurotransmission has been implicated in the pathogenesis of mood disorders such as major depressive disorder and a variety of other disorders. The 5-HT transporter protein (also called the 5-HTT) plays a key role in regulating the serotonergic system by regulating the reuptake of serotonin from the synaptic cleft following synaptic release. The 5-HTT protein is encoded by the serotonin transporter gene (*SLC6A4*). The serotonin transporter gene has a 44-bp insertion/deletion polymorphism in its regulatory region, known as the 5-HTT linked promoter region (5-HTTLPR). This deletion/insertion polymorphism results in two common alleles, the long allele (L) and the short allele (S). The S allele is associated with a 2- to 2.5-fold decrease in 5-HTT transcription rate compared with the L allele, resulting in less efficient serotonin reuptake and decreased expression of the serotonin transporter. The presence of the S allele has been reported to be associated with an increased susceptibility to major depression. Gressier et al. commented that the S allele (or SS genotype) seemed to be associated with an increased risk of depression, depressive symptoms, anxiety traits, and symptoms of internalizing behavior among women and an increased risk of aggressiveness, conduct disorder, and symptom counts of externalizing behavior among men. Moreover, the presence of stressful life events reinforced the association. Interestingly, these differences seemed to begin with adolescence and were not consistent among the elderly, suggesting a plausible role of hormonal fluctuations [87].

Covault combined daily reports of drinking and drug use obtained using a daily web-based survey with self-reports of past-year negative life events and *5-HTTLPR* genotypes in a regression analysis of alcohol and nonprescribed drug use in a sample of 295 college students. The authors observed that genotype and negative life events significantly interacted in relation to drinking and drug use outcomes. Individuals homozygous for the short (S) allele who experienced multiple negative life events in the prior year reported more frequent drinking and heavy drinking, stronger intentions to drink, and greater nonprescribed drug use. In individuals homozygous for the long (L) allele, drinking and drug use were unaffected by past-year negative life events. Heterozygous subjects showed drinking outcomes that were intermediate to the two homozygous groups. The authors concluded that the *5-HTTLPR* S allele is associated with increased drinking and drug use among college students who have experienced multiple negative life events. In addition, the S-allele carriers may be at risk for a variety of adverse behavioral outcomes in response to stress [88].

Association between μ-opioid receptor gene polymorphism (A118G) and opioid abuse has been discussed in detail earlier in the chapter. Genes encoding GABA receptors are of interest in search for genetic factors related to drug abuse. A dinucleotide repeat polymorphism of the GABA receptor β3 subunit gene (*GABARB3*) results in either the presence of 181 b.p-G1 or 11 other repeats designed as non-G1 alleles, including G2 (183 b.p.), G3 (185 b.p.), G4 (187 b.p.), G5 (189 b.p.), G6 (191 b.p.), G7 (193 b.p.), G8 (195 b.p.), G9 (197 b.p.), G10 (199 b.p.), G11 (201 b.p.), and G12 (203 b.p.). Nobel et al. reported that in the same population of nonalcoholics and alcoholics studied, variants of both the *DRD2* and *GABRB3* genes independently contribute to the risk for alcoholism, with the *DRD2* variants revealing a stronger effect than the *GABRB3* variants [89]. Feusner et al. observed an association between G1 heterozygosity in the GABA receptor β3 subunit gene and high levels of anxiety/insomnia, depression, somatic symptoms, and social dysfunction in a posttraumatic stress disorder population [90].

Monoamine oxidase A is a mitochondrial enzyme that is responsible for degradation of monoamines, including dopamine. The promotor region of the human *MAOA* gene which is located on the X chromosome has an upstream VNTR (*MAO-uVNTR*), including 2, 3, 3.5, 4, and 5 repeat of 30 bp. The allele with three repeats (allele 3) encodes monoamine oxidase A enzyme with lower activity, while allele with four repeats (allele 4) encodes enzyme with high activity. The carriers of allele 3 in males and the homozygotes of allele 3 in females are classified as the low activity group, the heterozygotes of alleles 3 and 4 in females as the medium activity group, and the carriers of allele 4 in males and the homozygotes of allele 4 in females as the high activity group. Shiraishi et al. based on one-way analysis of variance observed that the scores of novelty seeking and reward dependence were significantly higher in the high activity group than in the low activity group. The authors conclude that the monoamine oxidase A VNTR polymorphism that affects novelty seeking and reward dependence in healthy study participants with carriers of four repeats may have higher tendency of novelty seeking behavior [91]. Nakamura et al. commented that the high-activity allele class of *MAOA-u VNTR* in males may be involved in susceptibility to a persistent course of methamphetamine psychosis. However, authors found no difference in females [92]. *COMT* (val58Met) polymorphism has been discussed in detail earlier in the chapter.

EPIGENETIC ISSUES

Epigenetics is the regulation of the heritable and potentially reversible changes in gene expression that occur without altering the DNA sequence. Therefore, epigenetics is chemical modifications of either DNA or histone proteins that wrap DNA. The DNA-protein complex is known as chromatin. Histone chemical modifications take place mainly at lysine amino acid residue located in the tails of histones 3 and 4 (H3 and H4); however, other histones, such as H2A and H2B, can be modified as well. Covalent alterations, including acetylation, methylation, phosphorylation, ubiquitination, citrullination, and ADP-ribosylation, modulate the chromatin structure differently. Besides histone modifications, DNA methylation is a common epigenetic alteration capable of affecting the chromatin structure (DNA methylation is represented by the addition of a methyl group to the DNA sequence, in particular, to the cytosine nucleotides, and is regulated by DNA methyltransferases). When DNA is methylated, chromatin condenses and the transcription complex will not be able to bind DNA, thus inactivating the gene expression. Alcohol, as well as stress, can lead to epigenetic modifications that may be associated with synaptic remodeling and behavioral phenotypes such as anxiety and depression [93].

Epidemiological studies indicate that nongenetic factors contribute 40%–60% of the risk of developing drug addiction. Some of these are environmental and drug-induced factors, but other factors such as "epigenetic modifications" (i.e., DNA methylation and chromatin remodeling) may also play an important role. The transmission of information not encoded in the DNA sequence is termed epigenetic inheritance. DNA methylation and covalent histone modifications are the primary sources of epigenetic inheritance. Methylation of cytosine at the $5'$ position of the cytosine pyrimidine ring in CpG dinucleotide site (CpG represents regions of DNA where a cytosine nucleotide is followed by a guanine nucleotide; also known as CpG island) is a common epigenetic change. In general, methylation of the CpG sites in the promotor regions leads to decreased gene expression. In humans, there are about 45,000 CpG islands, many of which are found in the promoter regions of genes. These CpG islands are generally located upstream of the transcription start site to within the first exon. Approximately 70% of the CpG dinucleotides in the genome are methylated, while most of the CpG islands in the promoters of housekeeping genes (genes constitutively transcribed in most cells and representing 60% of the genome) are unmethylated. Genes without CpG islands, in general,

are repressed by the methylation of CpG dinucleotides in their promoter regions [94,95]. Nielsen et al. investigated whether there are differences in cytosine:guanine (CpG) dinucleotide methylation in the OPRM1 promoter region between former heroin addicts and controls. The authors analyzed methylation at 16 different CpG islands in DNA obtained from lymphocytes of 194 Caucasian former heroin addicts stabilized in methadone maintenance treatment and 135 Caucasian control subjects. Direct sequencing of bisulfite-treated DNA showed that the percent methylation at two CpG sites was significantly associated with heroin addiction. The level of methylation at the -18 CpG site was 25.4% in the stabilized methadone-maintained former heroin addicts and 21.4% in controls, and the level of methylation at the $+84$ CpG dinucleotide site was 7.4% in cases and 5.6% in controls. Both the -18 and the $+84$ CpG sites are located in potential Sp1 transcription factor–binding sites. Methylation of these CpG sites may lead to reduced OPRM1 gene expression in the lymphocytes of these former heroin addicts [96].

Egervari et al. used postmortem human brain specimens from a homogeneous European Caucasian population of heroin users for transcriptional and epigenetic profiling, as well as direct assessment of chromatin accessibility in the striatum, a brain region central to reward and emotion. A rat heroin self-administration model was used to obtain translational molecular and behavioral insights. The authors reported marked impairments related to glutamatergic neurotransmission and chromatin remodeling in the human striatum. The specific locations of epigenetic disturbances were hyper acetylation of lysine 27 of histone H3, showing dynamic correlations with heroin use history and acute opiate toxicology. Targeted investigation of GRIA1, a glutamatergic gene implicated in drug-seeking behavior, verified the increased enrichment of lysine 27–acetylated histone H3 at discrete loci, accompanied by enhanced chromatin accessibility at hyperacetylated regions in the gene body. Analogous epigenetic impairments were detected in the striatum of heroin self-administering rats. Overall, the authors demonstrated that the history of heroin use determines the degree of histone acetylation at specific lysine residues of the histone tail that links directly to impairments of glutamatergic genes involved in the regulation of synaptic plasticity in the striatum of human heroin users, establishing a pathological state that contributes to the continued cycles of drug-seeking behavior and relapse [97].

Drug-induced changes in gene expression in key brain reward regions, such as the NAc, PFC, and VTA, represent one mechanism by which epigenetic changes alter brain function in addicts. Acetylation of histone lysine residues reduces the electrostatic interaction between histone proteins and DNA, which is thought to relax chromatin structure and make DNA more accessible to transcriptional regulators. Histone acetylation is best characterized on histones H3 and H4: it can occur on lysines 9, 14, 18, and 23 on the N-terminal tail of H3 and at lysines 5, 8, 12, and 16 on the tail of H4. Genome-wide studies have shown that hyperacetylation in promoter regions is strongly associated with gene activation, whereas hypoacetylation causes reduced gene expression. This association also exists in the brain in vivo in response to drugs of abuse. Acute exposure to cocaine, for example, which is known to rapidly induce the immediate early genes c-fos and fosb in the NAc, increases histone H4 acetylation on their proximal gene promoters. Time course analysis revealed that this modification occurs within 30 min and disappears by 3 h, consistent with the induction kinetics of these immediate early genes. Therefore, acute cocaine does increase global levels of histone H4 acetylation and histone H3 phosphoacetylation but not H3 acetylation alone, within 30 min [98].

Browne et al. commented that the persistence of neuroadaptations due to opioid abuse is mediated in part by epigenetic remodeling of gene expression programs in discrete brain regions. Although the majority of work examining how epigenetic modifications contribute to addiction has focused on psychostimulants such as cocaine, more recently researchers are investigating epigenetic changes due to opioid abuse. Current evidence points toward opioids promoting higher levels of permissive histone acetylation and lower levels of repressive histone methylation, as well as alterations to DNA methylation patterns and noncoding RNA expression throughout the brain's reward circuitry. Additionally, studies manipulating epigenetic enzymes in specific brain regions are beginning to build causal links between these epigenetic modifications and changes in addiction-related behavior [99]. In-depth discussion on epigenetic changes due to drug abuse is beyond the scope of this chapter.

CONCLUSIONS

Both genetics and environment play important role in making and individual vulnerable to drug addiction. However no single gene has been linked to drug and or alcohol addiction. Instead, multiple genes contribute to overall susceptibility of an individual for drug and or alcohol abuse. The medical literature is also full of contradictory results with one group showing association between polymorphism of a particular gene with alcohol or drug abuse while another group indicating no such association. Moreover, epigenetic changes in genes may also be linked to susceptibility to alcohol and or drug abuse. Therefore, more investigations are needed to firmly establish genetic contribution to vulnerability of an individual to drug and or alcohol abuse.

REFERENCES

[1] American Psychiatric Association. Diagnostic and statistical manual of mental disorders, vol. 5. Washington, D.C: American Psychiatric Association; 2013.

[2] Venter JC, Adams MD, Myers EW, Li PW, et al. The sequence of human genome. Science 2001;291:1304—51.

[3] Prom-Wormley EC, Ebejer J, Dick DM, Bowers MS. The genetic epidemiology of substance use disorder: a review. Drug Alcohol Depend 2017;180:241—59.

[4] Goldman D, Oroszi G, Ducci F. The genetics of addictions: uncovering the genes. Nat Rev Genet 2005;6: 521—32.

[5] Johansson I, Ingelman-Sundberg M. Genetic polymorphism and toxicology-with emphasis on cytochrome P450. Toxicol Sci 2011;120:1—13.

[6] Zanger UM, Klein K. Pharmacogenetics of cytochrome P450 2B6 (CYP2B6): advances on polymorphisms, mechanisms, and clinical relevance. Front Genet 2013; 4:24.

[7] Wang B, Wang J, Huang SQ, Su HH, Zhou SF. Genetic polymorphism of the human cytochrome P450 2C9 gene and its clinical significance. Curr Drug Metabol 2009;10:781—834.

[8] Sienkiewicz-Oleszkiewicz B, Wiela-Hojeńska A. CYP2C19 polymorphism in relation to the pharmacotherapy optimization of commonly used drugs. Pharmazie 2018;73: 619—24.

[9] Werk AN, Cascorbi I. Functional gene variants of CYP3A4. Clin Pharmacol Ther 2014;96:340—8.

[10] Shi WL, Tang HL, Zhai SD. Effects of the CYP3A481b genetic polymorphism on the pharmacokinetics of tacrolimus on adult renal transplant recipients: a meta-analysis. PLoS One 2015;10:e0127995.

[11] Crews KR, Gaedigk A, Dunnenberger HM, Leeder JS, et al. Clinical Pharmacogenetics Implementation Consortium guidelines for cytochrome P450 2D6 genotype and codeine therapy: 2014 update. Clin Pharmacol Ther 2014; 95:376—82.

[12] Volpe DA, McMahon Tobin GA, Mellon RD, Katki AG, et al. Uniform assessment and ranking of opioid μ receptor binding constants for selected opioid drugs. Regul Toxicol Pharmacol 2011;59:385—90.

[13] Linares OA, Fudin J, Schiesser WE, Daly Linares AL, Boston RC. CYP2D6 phenotype-specific codeine

population pharmacokinetics. J Pain Palliat Care Pharmacother 2015;29:4−15.

[14] Kirchheiner J, Schmidt H, Tzvetkov M, Keulen JT, et al. Pharmacokinetics of codeine and its metabolite morphine in ultra-rapid metabolizers due to CYP2D6 duplication. Pharmacogenomics J 2007;7:257−65.

[15] VanderVaart S, Berger H, Sistonen J, Madadi P, et al. CYP2D6 polymorphisms and codeine analgesia in post-partum pain management: a pilot study. Ther Drug Monit 2011;33:425−32.

[16] Gasche Y, Daali Y, Fathi M, Chiappe A, et al. Codeine intoxication associated with ultrarapid CYP2D6 metabolism. N Engl J Med 2004;351:2827−31.

[17] Ciszkowski C, Madadi P, Phillips MS, Lauwers AE, Koren G. Codeine, ultrarapid-metabolism genotype, and postoperative death. N Engl J Med 2009;361: 827−8.

[18] Kelly LE, Rieder M, van den Anker J, et al. More codeine fatalities after tonsillectomy in North American children. Pediatrics 2012;129:e1343−1347.

[19] Virbalas J, Morrow BE, Reynolds D, Bent JP, Ow TJ. The prevalence of ultrarapid metabolizers of codeine in a diverse. Urban Population 2019;160:420−5.

[20] Madadi P, Koren G, Cairns J, Chitayat D, et al. Safety of codeine during breastfeeding: fatal morphine poisoning in the breastfed neonate of a mother prescribed codeine. Can Fam Physician 2007;53:33−5.

[21] Owusu Obeng A, Hamadeh I, Smith M. Review of opioid pharmacogenetics and considerations for pain management. Pharmacotherapy 2017;37:1105−21.

[22] Lassen D, Damkier P, Brosen K. The pharmacogenetics of tramadol. Clin Pharmacokinet 2015;54:825−36.

[23] Stamer UM, Lehnen K, Hothker F, Bayerer B, et al. Impact of CYP2D6 genotype on postoperative tramadol analgesia. Pain 2003;105:231−8.

[24] Wang G, Zhang H, He F, Feng X. Effect of the CYP2D6*10 C188T polymorphism on postoperative tramadol analgesia in a Chinese population. Eur J Clin Pharamcol 2006;62:927−31.

[25] Susce MT, Murray-Carmichael E, de Leon J. Response to hydrocodone, codeine and oxycodone in a CYP2D6 poor metabolizer. Prog Neuropsychopharmacol Biol Psychiatry 2006;30:1356−8.

[26] Orliaguet G, Hamza J, Couloigner V, Denoyelle F, et al. A case of respiratory depression in a child with ultrarapid CYP2D6 metabolism after tramadol. Pediatrics 2015; 135:e753−755.

[27] Stamer UM, Stuber F, Muders T, Musshoff F. Respiratory depression with tramadol in a patient with renal impairment and CYP2S6 gene duplication. Anesth Analg 2008; 107:926−9.

[28] Fortenberry M, Crowder J, So TY. The use of codeine and tramadol in the pediatric population-what is the verdict now? J Pediatr Health Care 2019;33:117−23.

[29] Samer CF, Daali Y, Wagner M, Hopfgartner G, et al. The effects of CYP2D6 and CYP3A activities on the pharmacokinetics of immediate release oxycodone. Br J Pharmacol 2010;160:907−18.

[30] Balyan R, Mecoli M, Venkatasubramanian R, Chidambaran V, et al. CYP2D6 pharmacogenetic and oxycodone pharmacokinetic association study in pediatric surgical patients. Pharmacogenomics 2017;18: 337−48.

[31] Klimas R, Witticke D, El Fallah S, Mikus G. Contribution of oxycodone and its metabolites to the overall analgesic effect after oxycodone administration. Expet Opin Drug Metabol Toxicol 2013;9:517−28.

[32] Hutchinson MR, Menelaou A, Foster DJ, Coller JK, Somogyi AA. CYP2D6 and CYP3A4 involvement in the primary oxidative metabolism of hydrocodone by human liver microsomes. Br J Clin Pharmacol 2004;57: 287−97.

[33] Madadi P, Hildebrandt D, Gong IY, Schwarz U, et al. Fatal hydrocodone overdose in a child: pharmacogenetics and drug interactions. Pediatrics 2010;126:e986−989.

[34] Li Y, Kantelip JP, Gerritsen-van-Schieveen P, Davani S. Interindividual variability of methadone response: impact of genetic polymorphism. Mol Diagn Ther 2008; 12:109−24.

[35] Kharasch ED, Regina KJ, Blood J, Friedel C. Methadone pharmacogenomics: CYP2B6 polymorphism determines plasma concentration, clearance and metabolism. Anesthesiology 2015;123:1142−53.

[36] Levran O, Peles E, Hamon S, Randesi M, et al. CYP2B6 SNPs are associated with methadone dose required for effective treatment of opioid addiction. Addiction Biol 2013;18:709−16.

[37] Bunten H, Liang WJ, Pounder DJ, Seneviratne C, et al. ORRM1 and CYP2B6 gene variants as risk factor in methadone-related deaths. Clin Pharmacol Ther 2010; 88:383−9.

[38] Madadi P, Kelly LE, Kepron C, Edwards JN, et al. Forensic investigation on methadone concentrations in deceased breastfed infants. J Forensic Sci 2016;61:576−80.

[39] Paul ABM, Simms L, Mahesan AM. The toxicology of methadone-related death in infants under 1 Year: three case series and review of the literature. J Forensic Sci 2017;62:1414−7.

[40] Mouly S, Bloch V, Peoch K, Houze P, et al. Methadone dose in heroin-dependent patients: role of clinical factors, comedication, genetic polymorphism and enzyme activity. Br J Clin Pharmacol 2015;79:967−77.

[41] Basatami S, Gupta A, Zackrisson AL, Ahlner J, et al. Influence of UGT2B7, OPM1 and ABCB1 gene polymorphism on postoperative morphine consumption. Basic Clin Pharmacol Toxicol 2014;115:423−31.

[42] Nielsen LM, Sverrisdóttir E, Stage TB, Feddersen S, et al. Lack of genetic association between OCT1, ABCB1, and UGT2B7 variants and morphine pharmacokinetics. Eur J Pharmaceut Sci 2017;99:337−42.

[43] Vandenbossche J, Richards H, Francke S, Van Den Bergh A, et al. The effect of UGT2B7*2 polymorphism on the pharmacokinetics of OROS® hydromorphone in Taiwanese subjects. J Clin Pharmacol 2014;54:1170−9.

[44] Xia S, Persaud S, Birnbaum A. Exploratory study on association of single-nucleotide polymorphisms with

hydromorphone analgesia in ED. Am J Emerg Med 2015; 33:444−7.

[45] Koolen SL, Van der Rijt CC. Is there a role for pharmacogenetics in the dosing of fentanyl? Pharmacogenomics 2017;18:417−9.

[46] Kuip EJ, Zandvliet ML, Koolen SL, Mathijssen RH, van der Rijt CC. A review of factors explaining variability in fentanyl pharmacokinetics; focus on implications for cancer patients. Br J Clin Pharmacol 2017;83:294−313.

[47] Hodges LM, Markova SM, Chinn LW, Gow JM, et al. Very important pharmacogene summary: ABCB1 (MDR1, P-glycoprotein). Pharmacogenetics Genom 2011;21: 152−61.

[48] Gray K, Adhikary SD, Janicki P. Pharmacogenomics of analgesics in anesthesia practice: a current update of literature. J Anaesthesiol Clin Pharmacol 2018;34: 155−60.

[49] Mamie C, Rebsamen MC, Morris MA, Morabia A. First evidence of a polygenic susceptibility to pain in a pediatric cohort. Anesth Analg 2013;116:170−7.

[50] Candiotti K, Yang Z, Xue L, Zhang Y, et al. Single-nucleotide polymorphism C3435T in the ABCB1 gene is associated with opioid consumption in postoperative pain. Pain Med 2013;14:1977−84.

[51] Park HJ, Shinn HK, Ryu SH, Lee HS, et al. Genetic polymorphisms in the ABCB1 gene and the effects of fentanyl in Koreans. Clin Pharmacol Ther 2007;81:539−46.

[52] Saiz-Rodríguez M, Ochoa D, Herrador C, Belmonte C, et al. Polymorphisms associated with fentanyl pharmacokinetics, pharmacodynamics and adverse effects. Basic Clin Pharmacol Toxicol 2019;124:321−9.

[53] Coulbault L, Beaussier M, Verstuyft C, et al. Environmental and genetic factors associated with morphine response in the postoperative period. Clin Pharmacol Ther 2006;79:316−24.

[54] Coller JK, Barratt DT, Dahlen K, Loennechen MH, Somogyi AA. ABCB1 genetic variability and methadone dosage requirements in opioid-dependent individuals. Clin Pharmacol Ther 2006;80:682−90.

[55] Dzambazovska-Trajkovska V, Nojkov J, Kartalov A, Kuzmanovska B, et al. Association of single-nucleotide polymorphism C3435T in the ABCB1 gene with opioid sensitivity in treatment of postoperative pain. Pril (Makedon Akad Nauk Umet Odd Med Nauki) 2016;37:73−80.

[56] Crist RC, Berrettini WH. Pharmacogenetics of OPRM1. Pharmacol Biochem Behav 2014;123:25−33.

[57] Yu Z, Wen L, Shen X, Zhang H. Effects of the OPRM1 A118G polymorphism (rs1799971) on opioid analgesia in cancer pain: a systematic review and meta-analysis. Clin J Pain 2019;35:77−86.

[58] Boswell MV, Stauble ME, Loyd GE, Langman L, et al. The role of hydromorphone and OPRM1 in postoperative pain relief with hydrocodone. Pain Physician 2013; 16(3):E227−35.

[59] Cajanus K, Kaunisto MA, Tallgren M, Jokela R, Kalso E. How much oxycodone is needed for adequate analgesia after breast cancer surgery: effect of the OPRM1 118A>G polymorphism. J Pain 2014;15:1248−56.

[60] Zwisler ST, Enggaard TP, Mikkelsen S, Verstuyft C, et al. Lack of association of OPRM1 and ABCB1 single-nucleotide polymorphisms to oxycodone response in postoperative pain. J Clin Pharmacol 2012;52:234−42.

[61] Haerian BS, Haerian MS. OPRM1 rs1799971 polymorphism and opioid dependence: evidence from a meta-analysis. Pharmacogenomics 2013;14.

[62] Chen D, Liu L, Xiao Y, Peng Y, et al. Ethnic-specific meta-analyses of association between the OPRM1 A118G polymorphism and alcohol dependence among Asians and Caucasians. Drug Alcohol Depend 2012;123:1−6.

[63] Türkan H, Karahalil B, Kadıoğlu E, Eren K, et al. The association between the OPRM1 A118G polymorphism and addiction in a Turkish population. Arh Hig Rada Toksikol 2019;70:97−103.

[64] Volkow ND, Morales M. The brain on drugs: from reward to addiction. Cell 2015;162:712−25.

[65] Oliva I, Wanat MJ. Ventral tegmental area afferents and drug dependent behavior. Front Psychiatr 2016;7:00030.

[66] Zhu F, Yan C, Wen YC, Wang J, et al. Dopamine D1 receptor gene variation modulates opioid dependance risk by affecting transition to addiction. PLoS One 2013;8: e70805.

[67] Liu JH, Zhong HJ, Dang J, Peng L, Zhu YS. Single-nucleotide polymorphisms in dopamine receptor D1 are associated with heroin dependence but not impulsive behavior. Genet Mol Res 2015;14:4041−50.

[68] Clarke TK, Weiss AR, Ferarro TN, Kampman KM, et al. The dopamine receptor D2 (DRD2) SNP rs1076560 is associated with opioid addiction. Ann Hum Genet 2014;78: 33−9.

[69] Khokhar JY, Ferguson CS, Zhu A, Tyndale RF. Pharmacogenetics of drug dependance: role of variation in susceptibility treatment. Annu Rev Pharmacol Toxicol 2010;50:39−61.

[70] Krebs MO, Sautel F, Bourdel MC, Sokoloff P, et al. Dopamine D3 receptor gene variants and substance abuse in schizophrenia. Mol Psychiatr 1998;3:337−41.

[71] Ray LA, Bryan A, Mackillop J, McGeary J, Hesterberg K, Hutchison KE. The dopamine D receptor (DRD4) gene exon III polymorphism, problematic alcohol use and novelty seeking: direct and mediated genetic effect. Addiction Biol 2009;14:238−44.

[72] Shao C, Li Y, Jiang K, Zhang D, et al. Dopamine D4 receptor polymorphism modulates cue-elicited heroin craving in Chinese. Psychopharmacology (Berlin) 2006;186: 185−90.

[73] Randesi M, van den Brink W, Levran O, Yuferov V, et al. Dopamine gene variants in opioid addiction: comparison of dependent patients, nondependent users and healthy controls. Pharmacogenomics 2018;19:95−104.

[74] Deng XD, Jiang H, Ma Y, Gao Q, et al. Association between DRD2/ANKK1 TaqIA polymorphism and common illicit drug dependence: evidence from a meta-analysis. Hum Immunol 2015;76:42−51.

[75] Xu K, Lichtermann D, Lipsky RH, Franke P, et al. Association of specific haplotypes of D2 dopamine receptor gene with vulnerability to heroin dependence in 2 distinct populations. Arch Gen Psychiatr 2004;61:597−606.

[76] Zhang F, Tong J, Hu J, Zhang H, Ouyang W, et al. COMT gene haplotypes are closely associated with postoperative fentanyl dose in patients. Anesth Analg 2015;120:933—40.

[77] Randesi M, van den Brink W, Levran O, Blanken P, et al. VMAT2 gene (SLC18A2) variants associated with a greater risk for developing opioid dependence. Pharmacogenomics 2019;20:331—41.

[78] Wu W, Zhu YS, Li SB. Polymorphisms in the glutamate decarboxylase 1 gene associated with heroin dependance. Biochem Biophys Res Commun 2012;422:91—6.

[79] Febo M, Blum K, Badgaiyan RD, Baron D, et al. Dopamine homeostasis: brain functional connectivity in reward deficiency syndrome. Front Biosci (Landmark Ed) 2017;22:669—91.

[80] Blum K, Oscar-Berman M, Barh D, Giordano J, Gold M. Dopamine genetics and function in food and substance abuse. J Genet Syndr Gene Ther 2013;4(121):1000121.

[81] Blum K, Oscar-Berman M, Giordano J, Downs B, et al. Neurogenetic impairments of brain reward circuitry links to reward deficiency syndrome (RDS): potential nutrigenomic induced dopaminergic activation. J Genet Syndr Gene Ther 2012;3(4).

[82] Liu HC, Chen CK, Leu SJ, Wu HT, Lin SK. Association between dopamine receptor D1 A-48G polymorphism and methamphetamine abuse. Psychiatr Clin Neurosci 2006; 60:226—31.

[83] Savitz J, Hodgkinson CA, Martin-Soelch C, Shen PH, et al. DRD2/ANKK1 Taq1A polymorphism (rs1800497) has opposing effects on D2/3 receptor binding in healthy controls and patients with major depressive disorder. Int J Neuropsychopharmacol 2013;16:2095—101.

[84] Bobadilla L, Vaske J, Asberg K. Dopamine receptor (D4) polymorphism is related to comorbidity between marijuana abuse and depression. Addict Behav 2013;38: 2555—62.

[85] Vasconcelos AC, Neto Ede S, Pinto GR, Yoshioka FK, et al. Association study of the SLC6A3 VNTR (DAT) and DRD2/ANKK1 Taq1A polymorphisms with alcohol dependence in a population from northeastern Brazil. Alcohol Clin Exp Res 2015;39:205—11.

[86] Gorwood P, Limosin F, Batel P, Hamon M, et al. The A9 allele of the dopamine transporter gene is associated with delirium tremens and alcohol-withdrawal seizure. Biol Psychiatr 2003;53:85—92.

[87] Gressier F, Calati R, Serretti A. 5-HTTLPR and gender differences in affective disorders: a systematic review. J Affect Disord 2016;190:193—207.

[88] Covault J, Tennen H, Armeli S, Conner TS, et al. Interactive effects of the serotonin transporter 5-HTTLPR polymorphism and stressful life events on college student drinking and drug use. Biol Psychiatr 2007;61:609—16.

[89] Noble EP, Zhang X, Ritchie T, Lawford BR, et al. D2 dopamine receptor and GABA(A) receptor beta3 subunit genes and alcoholism. Psychiatr Res 1998;81:133—47.

[90] Feusner J, Ritchie T, Lawford B, Young RM, et al. GABA(A) receptor beta 3 subunit gene and psychiatric morbidity in a post-traumatic stress disorder population. Psychiatr Res 2001;104:109—17.

[91] Shiraishi H, Suzuki A, Fukasawa T, Aoshima T, et al. Monoamine oxidase A gene promoter polymorphism affects novelty seeking and reward dependence in healthy study participants. Psychiatr Genet 2006;16:55—8.

[92] Nakamura K, Sekine Y, Takei N, Iwata Y, et al. An association study of monoamine oxidase A (MAOA) gene polymorphism in methamphetamine psychosis. Neurosci Lett 2009;455:120—3.

[93] Palmisano M, Pandey SC. Epigenetic mechanisms of alcoholism and stress-related disorders. Alcohol 2017;60: 7—18.

[94] Robertson KD, Wolffe AP. DNA methylation in health and disease. Nat Rev Genet 2000;1:11—9.

[95] Yuferov V1, Levran O, Proudnikov D, Nielsen DA, Kreek MJ. Search for genetic markers and functional variants involved in the development of opiate and cocaine addiction and treatment. Ann N Y Acad Sci 2010;1187: 184—207.

[96] Nielsen DA, Yuferov V, Hamon S, Jackson C, et al. Increased OPRM1 DNA methylation in lymphocytes of methadone-maintained former heroin addicts. Neuropsychopharmacology 2009;34:867—73.

[97] Egervari G, Landry J, Callens J, Fullard JF, et al. Striatal H3K27 acetylation linked to glutamatergic gene dysregulation in human heroin abusers holds promise as therapeutic target. Biol Psychiatr 2017;81:585—94.

[98] Kumar A, Choi KH, Renthal W, Tsankova NM, et al. Chromatin remodeling is a key mechanism underlying cocaine-induced plasticity in striatum. Neuron 2005;48: 303—14.

[99] Browne CJ, Godino A, Salery M, Nestler EJ. Epigenetic mechanisms of opioid addiction. Biol Psychiatr 2020; 87:22—33.

CHAPTER 5

Environmental Factors Linked to Opioids, Drugs, and Alcohol Abuse

INTRODUCTION

Opioid addiction is a serious, chronic, and relapsing psychiatric illness that could begin with recreational, experimental, and medical use. Although addiction can begin at any age, adolescents and young adults are at higher risk of experimenting with drugs of abuse. Unfortunately, opioid addiction can also develop after an individual is prescribed opioids for legitimate medical conditions [1]. It is estimated that 3.27%—11.5% of patients treated with opioids for chronic pain may develop chronic opioid addiction [2].

Fortunately, not all individuals using opioids will develop an addiction. It has been postulated that possibly a combination of genetic predispositions and environmental conditions together contribute to the risk of developing an opioid addiction. In general, multiple personality and environmental factors may increase the vulnerability of an individual to drug and or alcohol abuse. Moreover, multiple genetic factors may also increase the susceptibility of a person to substance abuse. In Chapter 4, genetic factors that may increase the risk of substance abuse is discussed in detail. In this chapter, the environmental factor including personality factors that may increase the vulnerability of a person to alcohol and drug abuse (collectively called substance abuse) is addressed.

Various personality factors such as novelty or sensation seeking, impulsivity, risk-taking behaviors, tolerance of deviance, and rebelliousness may increase the risk of substance abuse in a person. Sex differences may also play a role in determining the risk of drug abuse where male sex may be more susceptible to drug abuse. Studies demonstrated that initiation and escalation of drug use occur predominantly during adolescence and early adulthood. Moreover, experimentation with opioids during adolescence or early adulthood was demonstrated to be one of the major variables with the greatest correlation to adverse outcomes. Therefore, young adults should be counseled to avoid experimentation with drugs. Given this demographic, many studies examining the effects of social factors on drug abuse have focused on those particular age groups. Children of individuals abusing drugs are at higher risk of abusing opioids, compared to children of individuals not abusing drugs. Moreover, the social environment has a significant impact on drug use initiation especially drug use by parents, older siblings, and peers. More recently, the epidemic of opioid abuse resulted in studies that demonstrated similar roles for the immediate social group of adolescents and young adults in the abuse of prescription opioid medications [3,4].

SUBSTANCE ABUSE: A PSYCHIATRIC DISORDER

Behavioral addiction is defined as an intense desire to repeat some action that is pleasurable or perceived to improve well-being or capable of alleviating some personal distress, despite the awareness that such action may have negative consequences. From psychological, neurological, and social standpoint such repeated pattern of actions that are characterized as "addictive behavior" mimics drug as well as alcohol addiction. In the new "Diagnostics and Statistical Manual of Mental Disorders" (DSM-5; fifth edition published by American Psychiatric Association) gambling disorder is included in "Substance Related and Addictive Disorder" chapter because such behavior produces similar activation in the brain as seen with alcohol or substance use disorder [5].

Clinicians used DSM-IV criteria, which are structured interviews, for many years for a diagnosis of psychiatric illness including alcohol and substance use disorders that are psychiatric illness. There are significant overlaps between DSM-5 and DSM-IV criteria except DSM-IV describes two distinct disorders: alcohol abuse and alcohol dependence with specific criteria for each disorder. However, in DSM-5, both disorders are integrated into one disorder termed as alcohol use disorder (AUD) with subclassification of a mild, moderate, and severe disorder. There are 11 criteria in DSM-5 for the diagnosis of AUD and the presence of at least 2 of the

Fighting the Opioid Epidemic. https://doi.org/10.1016/B978-0-12-820075-9.00005-3

11 criteria in the past 12 months in a patient indicates that the patient is suffering from AUD. If only 2–3 criteria are present the diagnosis is mild, if 4–5 criteria are present then it is moderate and if 6 or more criteria are present then the diagnosis is severe AUD. In one study using DSM-5 criteria, the prevalence of AUD was 10.8% among 34,653 survey participants. According to DSM-IV criteria, 9.7% of the same population would have an AUD diagnosis [6]. DSM-5 is also used for the diagnosis of substance use disorder (cannabis, hallucinogens, inhalants, opioids, sedative/hypnotic, stimulants, and tobacco). In contrast to DSM-IV that has two criteria, substance abuse and substance dependence, DSM-5 guidelines combine them into one criterion: substance abuse disorder. There are other guidelines for the diagnosis of alcohol and or substance abuse disorder. For example, AUDIT (alcohol use disorder identification test) score can also be used to establish a diagnosis of alcohol dependence.

Grant et al. recently reported that based on DSM-5 criteria, the prevalence of lifetime drug use disorder in the United States was 9.9%. Drug use disorder included the use of sedatives/tranquilizers, cannabis, amphetamine, cocaine, heroin, and other opioids, hallucinogens, club drugs (MDMA, ketamine, etc.) as well as solvent/inhalant use. The highest prevalence of drugs were cannabis, opioids, and cocaine. The odds of drug use were generally greater among men, younger individuals who were unmarried or previously married, lower income, and those with high school and lower education. Moreover, white men and Native American individuals had a higher probability of drug abuse than men belonging to other ethnic groups. The drug use disorder was highly associated with alcohol use disorder and nicotine use disorder. The authors also observed a significant association between history of past 12 months drug use and major depressive disorder, dysthymia, bipolar disorder, posttraumatic stress syndrome, borderline personality disorder, antisocial personality disorder, and schizotypal personality disorder. Similar associations were observed among lifetime drug abusers. However, in addition to these disorders, associations between generalized anxiety disorders, panic disorders, social phobia, and drug use were also observed in lifetime drug users. Twelve-month past history of drug abuse was also associated with disability. Only 13.5% of respondents with a history of past 12 months drug use and only 24.6% of respondents with lifetime drug use received treatment indicating that drug use disorder is a common highly comorbid and disabling disorder that is largely untreated in the United States [7].

Many factors are associated with the susceptibility of a person to alcohol and or drug addiction. Certain environmental factors and personality traits may precipitate the risk of alcohol and/or drug addiction in certain individuals. In many instances, an individual may abuse both alcohol and drugs. In general, more men than women abuse alcohol and/or drugs. Women typically initiate drug abuse later than men, and their drug abuse is often influenced by the boyfriend of a spouse. Moreover, women and men use drugs for different reasons but women often enter treatment earlier than men. However, women also have a higher prevalence of comorbid psychiatric disorders such as depression and anxiety than men, and such mood disorders are often associated with substance abuse. Although such factors may appear to complicate treatment in women, research indicates that men and women are equally responsive to drug rehabilitation treatment [8].

FETAL EXPOSURE TO OPIOIDS

Infants exposed to tobacco, alcohol, prescription medications, and illicit substances in utero may exhibit signs of withdrawal from these substances after birth. Neonatal abstinence syndrome is a broad, nonspecific term assigned to this type of presentation in the newborn. Neonatal abstinence syndrome is a result of the sudden discontinuation of fetal exposure to substances that were used or abused by the mother during pregnancy. Withdrawal from licit or illicit substances is becoming more common among neonates in both developed and developing countries. The pathophysiology of neonatal abstinence syndrome is not completely understood. Urine or meconium confirmation may assist the diagnosis and management of such syndrome. The Finnegan scoring system is commonly used to assess the severity of neonatal abstinence syndrome; scoring can be helpful for initiating, monitoring, and terminating treatment in neonates. Nonpharmacological care is the initial treatment option, and pharmacological treatment is required if an improvement is not observed after nonpharmacological measures or if the infant develops severe withdrawal symptoms. Morphine is the most commonly used drug in the treatment of neonatal abstinence syndrome. Breastfeeding is not contraindicated, unless the mother is taking street drugs, is involved in polydrug abuse, or is infected with HIV [9].

However, the more specific term neonatal opioid withdrawal syndrome is becoming more widely used because many pregnant women abuse opioids. The rising incidence of maternal opioid use is demonstrated by hospital discharge records revealing a nationwide increase from 1.19 to 5.63 per 1000 births per year from 2000 to 2009. In 2012, an estimated 5.9% of women aged 15–44 years were using illicit drugs during pregnancy. Although abuse of marijuana was most common, prescription opioids abuse ranked number two after marijuana abuse. Because of opioid abuse by pregnant women in all parts of the United States as well among wide demographic characteristics, all healthcare providers must be aware of the management of neonatal opioid withdrawal. Mothers who are abusing opioids may be identified during prenatal care and referred to perinatal substance abuse programs, which will optimally have an affiliated neonatal program. Unfortunately, many women with opioid addiction do not obtain prenatal care and are first seen when they present in labor. Some women, particularly those with addiction to prescription opioids, may be able to obtain a prescription from other physicians or purchase diverted "street drugs" during the pregnancy. As a result, neonatal exposure will be unsuspected until the withdrawal syndrome develops after birth. Risk factors for maternal drug abuse include poor or no prenatal care, a previously unexplained late fetal demise placental abruption, unexplained intrauterine growth restriction, maternal hypertension, and precipitous labor. These factors, as well as clinical suspicion of opioid withdrawal, or a known history of maternal drug abuse or opioid replacement therapy may prompt screening with a maternal or neonatal urine drug screen or meconium toxicology testing. The legal implications of this screening are important to consider before initiation, as several states consider a positive newborn urine drug screen to be evidence of child abuse. The optimal urine sample for neonatal screening is the first urine after birth, as many substances are quickly metabolized and become undetectable. It has been estimated that 21%–94% of neonates exposed to opioids in utero will develop withdrawal signs and symptoms that are severe enough to warrant pharmacologic treatment [10].

According to a recent survey, only about half of neonatal intensive care units in the United States have a written protocol for the diagnosis and management of neonatal withdrawal symptoms, which represents an important area for educational improvement. It has been estimated that approximately 5.8 of 1000 pregnancies in the United States are complicated by the use of opioids, although rates vary widely by region [11]. Although women who abuse opioids during pregnancy can be treated with methadone or buprenorphine, there are many complications of opioid abuse during pregnancy both to the mother and the newborn [12].

Opioids use during pregnancy is associated with higher rates of depression, anxiety, and chronic medical conditions. Whiteman et al. reported that after adjusting for confounders, opioid use was associated with increased odds of threatened preterm labor, early-onset delivery, poor fetal growth, and stillbirth. Opioid users were four times as likely to have a prolonged hospital stay during labor and delivery. The authors also observed that pregnant women who used or abused opioids during pregnancy were more likely to have other comorbidities, including depression, anxiety, insomnia, diabetes, hypertension, renal diseases, and HIV infection. The mean per-hospitalization cost of a woman who used opioids during pregnancy was $5616 compared to $4084 for nonusers. The authors concluded that opioid use during pregnancy is associated with adverse perinatal outcomes and increased healthcare costs [13].

Pregnancy in opioid users is a critical clinical issue. Most guidelines recommend maintenance treatment to manage opioid addiction in pregnancy, with methadone being the gold standard. More recently, buprenorphine has been discussed as an alternate medication. The use and efficacy of buprenorphine in pregnancy are still controversial. For maternal outcomes, most studies suggest that buprenorphine has similar effects on methadone. However, neonatal abstinence syndrome is common in infants of both buprenorphine- and methadone-maintained mothers. As regards to neonatal outcomes, buprenorphine has the same clinical outcome as methadone, although some newer studies suggest that it causes fewer withdrawal symptoms [14]. Tran et al. commented that both methadone and buprenorphine are widely used to treat opioid use disorder; however, compared with methadone, buprenorphine is associated with shorter treatment duration, less medication needed to treat neonatal abstinence syndrome symptoms, and shorter hospitalizations for neonates. Naltrexone is not a first-line treatment for opioid abuse in pregnancy primarily because both detoxification and an opioid-free period are required [15].

Opioid Abuse in Pregnancy and Clinical Outcome

The potential teratogenic effects of maternal opioid use during pregnancy are also an area of great public health concern. Congenital malformations are serious, often costly medical conditions that can cause lifelong challenges. Opioid abuse during pregnancy is a leading cause of infant death in the United States, accounting for 20% of all deaths during the first year of life [16]. Congenital malformations can occur at any time during pregnancy, but the first trimester is typically the most vulnerable period. Two studies funded by the Centers for Disease Control and Prevention have linked opioid use during early pregnancy to congenital malformations. These studies report a twofold increased risk for some congenital heart defects, neural tube defects, and gastroschisis [17,18]. Lind et al. performed a systematic review of studies investigating the link between maternal opioid abuse and congenital malformations in neonates. Among the case-control studies, associations with oral clefts and ventricular septal defects/atrial septal defects were the most frequently reported specific malformations in neonates born to mothers abusing opioids during pregnancy [19].

Garrison et al. relied on the data obtained from two prospective cohorts at the University of New Mexico. For both cohorts, pregnant women were recruited during one of their prenatal care visits and followed up to delivery. The merged sample included 59 polydrug users, 22 exclusive opioid users, and 278 abstinent controls. Continuous growth measures (birth weight, height, occipital frontal circumference (OFC), and corresponding sex-specific percentiles) were compared by ANOVA and ANCOVA in bivariate and multivariable analyses, respectively. Based on data analysis the authors concluded that the risk of microcephaly (OFC < 10th percentile) was significantly greater in the polydrug abuser but exclusive for opioid groups compared to abstinent controls [20].

Many studies have demonstrated the harmful effects on the fetus/child of maternal abuse of opiates or opioids during pregnancy, and when taken during the late part of pregnancy neonatal abstinence symptoms are common [21]. Tramadol is a commonly prescribed and centrally acting atypical opioid analgesic. In one study, using the Swedish Medical Birth Register, the authors identified women (deliveries in 1997–2013) who had reported the use of tramadol in early pregnancy. Among 1,682,846 women (1,797,678 infants), 1751 women (1776 infants) had used tramadol. The authors reported that 96 of the infants had a congenital malformation and 70 of them were relatively severe. The adjusted odds ratio for a relatively severe malformation was 1.33. The odds ratio for cardiovascular defects was 1.56 and for pes equinovarus the odds ratio was significantly increased to 3.63. The study suggests a teratogenic effect of tramadol [22]. Therefore, chronic opioid therapy during pregnancy is perilous [23].

Disturbances of growth and behavior in infants and toddlers of women addicted to heroin during pregnancy have been reported in uncontrolled studies. In a controlled study by Wilson et al., 3–6-year-old children

of heroin-addicted mothers were compared to three other groups matched for age, race, sex, birth weight, and socioeconomic status. Heroin-exposed children weighed less and were shorter than those in the comparison groups: 14% had a head circumference below the third percentile. Heroin-exposed children were rated by parents as less well-adjusted than control children and they differed significantly in perceptual measures and on subtests of the Illinois Test of Psycholinguistic Abilities and McCarthy Scales of Children's Abilities relating to the process of the organization. The authors concluded that chronic intrauterine exposure to heroin may affect growth and behavior as well as perceptual and learning processes in preschool children [24].

Nygaard et al. assessed the cognitive functioning of 72 children with prenatal opioid and polysubstance exposure and 58 children without any established prenatal risk at 1, 2, 3, 4½, and 8½ years old and observed that the exposed boys had significantly lower levels of cognitive functioning than the control group, whereas there were increasing differences over time for the girls. The children in the exposed group had significantly lower IQ scores than the control group on the Wechsler Intelligence Scale for Children. The authors concluded that while effects of prenatal substance exposure cannot be isolated, group effects on cognition rather increased than waned over time, even in adoptive/foster children with minimal postnatal risk [25]. However, buprenorphine treatment of maternal opioid use disorder during pregnancy was not associated with greater harms than methadone treatment, and moderately strong evidence indicated a lower risk of preterm birth, greater birth weight, and larger head circumference with buprenorphine treatment compared to methadone treatment [26].

CHILDHOOD MALTREATMENT AND INCREASED RISK OF ADDICTION

The Federal Child Abuse Prevention and Treatment Act defines maltreatment as child abuse or neglect that encompasses any act or lack of action by a child's caretaker that results in physical or emotional harm. An extensive number of risk factors may contribute to the onset of substance use among adolescents. Familial risk factors include childhood maltreatment (including abuse and neglect), parental or familial substance abuse, marital status of parents, level of parental education, parent–child relationships, familial socioeconomic status, and child perception that parents approve of their substance use. Childhood maltreatment, including physical abuse and neglect, has been linked to increased risk for adolescent substance use, with one study reporting 29% of children who experienced maltreatment participating in some level of substance use [27] and another reporting 16% of maltreated children abusing drugs [28]. Being a victim of physical or sexual assault increases the risk of an adolescent getting involved with drug abuse from two to four times. However, different studies have shown varying specific results regarding which type of childhood abuse is the strongest contributor, with some reporting a higher risk associated with sexual abuse but others report a higher risk associated with physical abuse. Posttraumatic stress disorder (PTSD) is also associated with an increased likelihood of developing a substance use disorder, particularly with marijuana or hard drugs (including LSD, cocaine, heroin, inhalants, and nonmedical prescription drugs). This increased risk may be a result of the fact that trauma that typically leads to PTSD is highly stressful and may lead PTSD sufferers to cope with intense stress through substance use [29].

Males are more likely to be physically abused, whereas females are generally more likely to be sexually abused. However, generally speaking, gender differences regarding substance use vary widely across the literature. Age, though, shows consistent patterns, with older adolescents participating in substance use more often than their younger counterparts, with risk increasing each year from ages 10 to 17. One review of 35 studies indicated that most findings consistently show that childhood maltreatment is a risk factor for earlier onset of substance use [30]. This may be because victims of maltreatment use drugs and alcohol as coping mechanisms rather than purely for social reasons [31].

Physical abuse during childhood is associated with a higher risk of adolescent alcohol or drug abuse. In most states, the definition of physical child abuse is any act that causes a child physical harm and such harm is not accidental. Although boys are more likely to get physically abused girls often experience sexual abuse. Many studies have shown that physical abuse and sexual abuse predicted later substance abuse. In one study based on a clinical sample of 655 adolescents divided into two groups (polydrug users and nonpolydrug users), the authors observed that polydrug abusers had a greater prevalence of all types of maltreatment (physical abuse, sexual abuse, emotional abuse, physical neglect, and emotional neglect) although most associated with this group were sexual abuse and emotional neglect. Other relevant variables to adolescent drug abuse were diagnosis of depressive disorder, the presence of anxiety trait, and family history of alcohol dependence [32]. Skinner et al. followed 332

participants from childhood (18 months–6 years old) to adulthood (31–41 years old) and observed that childhood sexual abuse was fully mediated by adolescent alcohol use and depression. Children exposed to severe emotional abuse are at high risk of comorbid substance abuse, depression, and anxiety in their adolescent years and also mid-30s [33].

A variety of evidence demonstrated that attachment between parents and children plays a crucial role in healthy development and, in contrast, impaired parental bonding seems to be a major risk factor for mental diseases including substance abuse disorder [34]. Nonmedical prescription opioid use and childhood maltreatment are currently serious problems among adolescents worldwide, and childhood maltreatment may be associated with the increased rates of nonprescription opioid use among adolescents. Lei et al. investigated possible associations between particular types of childhood maltreatment and opioid use and assessed whether gender has a moderating effect on these associations in a Chinese population. In general, childhood maltreatment includes sexual abuse and all forms of physical and emotional ill-treatment and neglects that results in significant behavior problems as well as increased risk of addiction later in life. Childhood maltreatment is a global issue affecting more than two billion children worldwide, the majority of the children live in low- and middle-income countries. In China, childhood maltreatment is a common experience for children and is associated with huge financial losses because many maltreated children suffer psychological distress and might adopt unhealthy behaviors that increase the risk of chronic diseases. Based on data from the School-based Chinese Adolescents Health Survey ($n = 10,904\%$; 47.1% were male, and 52.9% were female; age range: 12–20 years), the authors observed that all five types of childhood maltreatment including physical abuse, emotional abuse, sexual abuse, physical neglect, and emotional neglect were positively associated with lifetime nonmedical opioid abuse. The female students who reported having experienced emotional/sexual abuse had a significantly higher prevalence of lifetime nonmedical opioid use than the male students [35].

Self-reported childhood maltreatment is also common in the United States. In one study based on self-reported 10,828 adolescents in Grade 7–12, the authors reported that supervision neglect was most common, (41.5%) followed by physical assault (28.4%). Half of those respondents who reported any physical assault, or 14.2% of all respondents, said this happened ≥3 times. The third most prevalent type of maltreatment was physical neglect, where more than 1 in 10 respondents (11.8%) recalled ≥1 episode of physical neglect, and 1 in 20 respondents (5.0%) reported ≥3 episodes. Finally, by the time they entered the sixth grade, more than 1 in 25 respondents (4.5%) said that they had been victims of contact sexual abuse perpetrated by a parent or other adult caregiver [36]. These results are summarized in Table 5.1.

In another study based on 12,388 subjects (male: 5604, female: 6684, most of the analytic samples were white, approximately 65%, and three-quarters had more than high school education), the authors reported that prevalence of the individual exposures ranged from 5.13% (experienced violence) to 16.37% (emotional abuse). Females relative to males had a somewhat higher prevalence of emotional abuse (19.37% vs. 13.43%) and sexual abuse (9.82% vs. 6.68%). The authors observed a correlation between the misuse of prescription pain medications and childhood abuse/trauma [37].

Barahmand et al. investigated the relative mediating effects of impulsivity and emotion dysregulation in the relationship between childhood maltreatment and motives for opiate use. The authors recruited 74 adolescent users of tramadol from a boot camp for deaddiction and rehabilitation services for the study. Participants completed assessments of childhood abuse history, difficulties regulating emotions, impulsiveness, and motives for substance use as well as a sociodemographic information sheet. The authors observed that emotion dysregulation, but not impulsiveness, mediated the relationship between childhood emotional abuse and expansion and enhancement motives for substance use. Therefore, difficulties regulating emotions may function as a mechanism linking prior childhood experiences of emotional abuse to subsequent motives for substance abuse. The authors suggested that targeting emotion dysregulation problems may be an effective adjunct in the treatment of childhood emotional abuse adolescent victims at risk for substance use [38]. Wolff

TABLE 5.1
Childhood Maltreatment in the United States.

Type of Neglect	Definition	Any Report (% Respondent)	Reported Once (% Respondent)	Reported Twice (% Respondent)	Reported Thrice or More Times (% Respondent)
Supervision neglect	Parents or other adult caregivers left the child home alone when an adult should have been supervising them	4184 (41.5%)	1276 (13.1%)	942 (9.3%)	1966 (19.1%)
Physical assault	Defined as being "slapped, hit, or kicked" by a parent or other adult caregiver	3013 (28.4%)	838 (8.3%)	649 (5.9%)	1526 (14.2%)
Physical neglect	Defined as an occasion where a parent or other adult caregiver did not meet their child's basic needs, such as keeping him or her clean or providing food or clothing.	1205 (11.8%)	471 (11.8%)	217 (2.1%)	517 (5.0%)
Contact sexual abuse	Sexual abuse by a parent or other adult caregiver	479 (4.5%)	233 (2.4%)	76 (0.5%)	170 (1.6%)

Source of data: Mills R, Scott J, Alati R, O'Callaghan M; Najman J, Strathearn L. Child maltreatment and adolescent mental health problems in a large birth cohort. Child Abuse Negl. 2013;37:292–302.

et al. commented that emotion dysregulation is a potential mechanism underlying the relationship between early emotional and physical maltreatment and the development of substance use disorder. Therefore, focusing on the early training of adaptive emotion regulation strategies after childhood maltreatment might be of considerable relevance to prevent the development of substance use disorder [39].

Child maltreatment has been frequently identified in the life histories of adolescents and adults in treatment for substance use disorders as well as in epidemiological studies of risk factors for substance use and substance use disorder [40]. It has been documented that childhood emotional maltreatment, physical maltreatment, and sexual abuse are associated with increased risk for tobacco use, alcohol use, illicit drug use, and polydrug use. More disturbingly, adolescent girls with a history of childhood sexual abuse are approximately five times more likely to be heavy polysubstance users compared to adolescent girls without experiences of childhood sexual abuse. Unfortunately, in comparison to other forms of childhood adversities (e.g., parental death, parental incarceration, parental divorce), child maltreatment uniquely predicted persistent alcohol dependence in adulthood [41]. Importantly, child maltreatment often occurs within the context of parental substance use disorder. Results of the Adverse Childhood Experiences study indicated that adults who reported being raised by one or more alcoholic parents were twice as likely to also have experienced child maltreatment [42]. In addition, prospective studies show that parental substance use disorder is a robust risk factor for child maltreatment [43]. Because individuals who experienced child maltreatment are at heightened risk for the development of substance use disorder, prevention efforts aimed at the avoidance of maltreatment are very important [44].

Taplin et al. based on a study of 87 subjects mean age 38 years concluded that maternal alcohol and drug use were significantly associated with childhood sexual abuse, emotional abuse, and physical neglect. Paternal alcohol and drug use were significantly associated with childhood physical abuse. Increased severity of all types of childhood trauma was related to an earlier age of the first injection of illicit drugs [45].

Stressful environmental factors including emotional neglect as a child could affect individual personality traits and mental health, possibly inducing stable changes in the hypothalamic–pituitary–adrenal (HPA) axis and brain mono-amine function, which in turn are involved in vulnerability to addictive behavior. Based on a study of 50 abstinent cocaine-dependent patients, and 44

normal controls, matched for age and sex, Gerra et al. observed that cocaine-addicted individuals, in general, showed significantly lower homovanillic acid (HVA), and higher prolactin, adrenocorticotropic hormone (ACTH), and cortisol-basal levels with respect to controls. In particular, neuroendocrine changes characterized cocaine addicts with a childhood history of neglect and low perception of parental care. Obsessive-compulsive, depression, and aggressiveness symptoms are related to poor parenting, inversely associated with HVA levels and directly associated with prolactin, ACTH, and cortisol levels. The authors suggest the possibility that childhood experience of neglect and poor parent–child attachment may partially contribute to a complex neurobiological derangement including HPA (hypothalamic–pituitary–adrenal) axis and dopamine system dysfunctions, playing a crucial role in addictive and affective disorders susceptibility [46].

Although childhood sex abuse is less common than other childhood maltreatment, sex abuse has profound ill effects when the child becomes an adolescent and then an adult. Children who have exposure to child sexual abuse are at particular risk for developing substance abuse in adolescence. Shin et al. examined the association between child sexual abuse and qualitatively distinct patterns of adolescent substance use by studying 1019 adolescents (mean age: 15.9 years; range: 13–18) who were selected from five publicly funded service systems. The authors reported that child sexual abuse was associated with an increased risk of being a heavy polysubstance user in girls, even after adjustment of age, race/ethnicity, parental substance use, sibling use, peer use, psychopathology and other forms of childhood maltreatment including physical abuse and neglect. From a clinical perspective, the most troublesome group was the polysubstance abusers where approximately 100% have used alcohol, cannabis, amphetamines, and cocaine. In addition, approximately 25% also reported the use of opioids and 82% hallucinogens [47]. Based on a study of 51 outpatient women with opioid use disorder who were on opioid maintenance treatment with buprenorphine/naloxone and 48 women in the control group with no history of opioid abuse, the authors concluded that childhood sexual abuse was associated with substance abuse disorder later in life [48].

Heffeman et al. investigated associations between childhood sexual and/or physical abuse and adult substance abuse in general by studying 763 consecutively admitted psychiatric inpatients in their hospital. Patients were interviewed about demographic information, alcohol and drug use, and a history of interpersonal violence. The authors reported that opiate users were 2.7 times more likely to have a history of childhood sexual and/or physical abuse than nonopiate users, after controlling for diagnostic and sociodemographic variables. Opiate use was higher among those reporting physical abuse alone (24.1%) or both physical and sexual abuse (27%) than among those reporting sexual abuse alone (8.8%) [49]. In another study, the authors examined the prevalence of five types of childhood trauma among a sample of adult patients who were addicted to opioids and seeking treatment with buprenorphine. Using a survey methodology, the authors examined a consecutive sample of 113 participants and found that 20.4% reported having experienced sexual abuse, 39.8% reported having experienced physical abuse, 60.2% reported having experienced emotional abuse, 23.0% reported having experienced physical neglect, and 65.5% reported having witnessed violence. Only 19.5% of the sample denied having experienced any of the five forms of childhood trauma. Most respondents (60.2%) reported having experienced one, two, or three different forms of childhood trauma. A minority reported having experienced four (13.3%) or all five (7.1%) forms of childhood trauma. These data indicate that among individuals with opioid dependence who are seeking treatment with buprenorphine, the prevalence rates of various types of childhood trauma are quite high [50].

ENVIRONMENTAL RISK FACTORS FOR ADOLESCENCE

Family studies have shown that the risk for alcohol dependence is 4−10-fold higher in the offspring of an alcoholic parent due to combined genetic and environmental factors. Children of alcoholics are considered to be at high risk because there is a greater likelihood that they will develop alcoholism compared with a randomly selected child from the same community. Children of alcoholics and children of drug-abusing parents are especially vulnerable to the risk of maladaptive behavior because they have combinations of many risk factors present in their lives. However, the most important risk factor is their parent's substance-abusing behavior that may put children of substance abusers at biologic, psychological, and environmental risk. A series of studies measured mortality, physiology, and general health in the offspring of alcoholic parents and concluded that when mothers stopped drinking during gestation, their children were healthier. Today, research can be classified into studies of fetal alcohol syndrome, the transmission of alcoholism, psychobiologic markers of vulnerability, and psychosocial characteristics. Each of these studies hypothesizes that differences between children of alcoholics and children of nonalcoholics influence maladaptive behaviors later in life, such as academic failure or alcoholism. In contrast, relatively little is known about children of heroin addicts, cocaine abusers, or polydrug abusers. Nonetheless, many researchers suggest that the children of addicted parents are at greater risk for later dysfunctional behaviors including substance use disorder [51].

Many environmental household factors are linked with the risk of substance abuse in adolescent years. Jurich et al. concluded that nine family factors that have an impact on drug abuse include parental absence, discipline, scapegoating, hypocritical morality, parent-child communication gap, parental divorce, mother-father conflicts, family breakups, and the use of "psychological crutches" to cope with stress. A psychological crutch in parallel to a physical crutch is anything a person relies on during a vulnerable situation. Therefore, a psychological crutch could be chemical dependence, emotional dependence on others and even physical dependence, for example, shopping or aerobic exercise with others when a person is going through a very difficult time such as divorce, death in a family, and losing a job [52].

Tomcikova et al. explored the associations between family composition, the quality of adolescents' communication with parents and adolescents' recent frequent alcohol drinking and lifetime drunkenness. The authors derived their data from the Slovak part of the 2005−06 Health Behavior in School-aged Children (HBSC) study. The sample consisted of 3882 students (46.3% males; mean age 13.3 years). Data on drinking alcohol in the past week, lifetime drunkenness, communication, and family composition were collected via anonymous questionnaires stratified for ages 11, 13, and 15 years. The authors reported that living in a split family increased the risk of frequent drinking and drunkenness among adolescents as well as a low quality of communication between mothers and their children. Risks were higher for drunkenness than for frequent alcohol use and strongly increased by age, with the communication with parents worsening at increasing age. The authors commented on the importance of quality of communication between parents and adolescents in preventing hazardous alcohol use among adolescents. Preventive interventions to reduce adolescents' use of alcohol should therefore also target the quality of communication in the family [53].

Behavioral factors such as excessive alcohol consumption play a major role in the explanation of social inequalities in health. The unequal distribution of health risk behaviors among socioeconomic groups has important consequences for both the current and future health status of the younger generation. Richter et al. investigated socioeconomic differences in adolescent drinking behavior among 11–15-year old adolescents in Europe and North America by using data obtained from the HBSC study 2001/02, a cross-national survey conducted in collaboration with the World Health Organization. The study included 69,249 male and 73,619 female students from 28 countries. The effect of parental occupation and family affluence on episodes of drunkenness was assessed using separate logistic regression models controlling for age. The authors reported that the socioeconomic circumstances of the family had only a limited effect on repeated drunkenness in adolescence. For girls only in 1 out of 28 countries a significant association between family affluence and repeated drunkenness was observed, while boys from low and/or medium affluent families in nine countries faced a lower risk of drunkenness than boys from more affluent families. Regarding parental occupation, significant differences in episodes of drunkenness were found in nine countries for boys and in six countries for girls. Compared to family affluence, which was positively related to the risk of drunkenness, a decreasing occupational status predicted an increased risk of drunkenness. This pattern was identified within several countries, most noticeably for boys. The authors concluded that parental socioeconomic status is only of limited importance for episodes of drunkenness in early adolescence, and this very limited role seems to apply for girls more than for boys and for parental occupation more than family affluence [54].

Melotti et al. asked a total of 5837 children aged 13 years about the previous consumption of alcohol and tobacco. Information on parental socioeconomic position, collected from questionnaires from the mother, included both social class and education of the mother and her partner and average household disposable income in early preschool childhood. The authors reported that consuming alcoholic drinks in the previous 6 months was linearly associated with higher income levels even when adjusting for other socioeconomic indicators. In contrast, both the risk of binge drinking and recent drinking was lower for children whose mothers had a higher educational level. Smoking tobacco was associated with lower socioeconomic position irrespective of the indicator used. The authors concluded that alcohol drinking was more common in young people from higher-income households but

less common with higher levels of maternal education. A consistent inverse socioeconomic gradient with tobacco smoking was apparent. These results may reflect how different aspects of socioeconomic position can influence health behavior in opposing directions. Higher income may increase the availability of alcohol in the family, whereas mothers with higher educational attainment might encourage more healthy behaviors in their children, including reduced alcohol use [55].

In one study, the authors explored the role of having rules about alcohol use, parental norms about early alcohol use, and parental alcohol use in the development of adolescents' drinking behavior. The author collected longitudinal data from 416 families consisting of both parents and two adolescents (aged 13–16 years). The analysis using structural equation modeling indicated that having clear rules decreases the likelihood of drinking in adolescence. However, longitudinally alcohol-specific rules have only an indirect effect on adolescents' alcohol use, namely through earlier drinking. Analyses focusing on explaining the onset of drinking revealed that having strict rules was related to the postponement of drinking initiation of older and younger adolescents. The author concluded that having strict rules is related to the postponement of drinking and that having alcohol-specific rules depends on other factors, thus underlining the complexity of the influence of parenting on the development of adolescents' alcohol. However, parents who drink alcohol may consider themselves less credible to impose strict drinking rules. Therefore, parent's permissiveness toward drinking is associated with a higher frequency of heavy alcohol consumption by adolescents [56].

Intimate Partner Violence and Addiction in Adolescents

Intimate partner violence is defined as any behavior within an intimate relationship (married, unmarried, and live-in) that causes physical, psychological, or sexual harm to those in that relationship. This definition encompasses physical, sexual, and psychological aggression/abuse or controlling the behavior of any kind. Domestic violence is defined as the physical, sexual, and emotional maltreatment of one family member by another. Domestic violence typically includes all types of family violence such as elder abuse, child abuse, and marital rape but intimate partner violence is limited to acts of aggression between intimate partners. Females are more likely to get hurt in intimate partner violence. UNICEF (The United Nations Children's Fund, originally known as the United Nations International Children's Emergency Fund) has estimated that between 133 million to 275 million children around the world witness frequent parental intimate partner conflict or violence. Studies have shown that 1 in 4 women and 1 in 7 men in the United States had experienced intimate partner violence in their lifetime. Schiff reported that women who experienced intimate partner conflict were 4.43 times more likely to develop depression later in life and who experienced intimate partner violence were 7.64 times more likely to develop depression. For young males who were exposed to parental intimate partner violence at 14 years old were more likely to develop anxiety, nicotine, alcohol, and cannabis use disorder later in life, but in general female offspring were more affected than males [57].

Smith et al. used a subsample ($n = 508$) of participants from the Rochester Youth Development Study, a longitudinal study of urban, largely minority adolescents that surveyed youth at high risk for antisocial behavior and drug use, in their investigation. The authors used logistic regression analyses to assess whether adolescent exposure to intimate partner violence predicted increased odds of four indicators of problem substance use in early adulthood, controlling for parental substance use, adolescent maltreatment, and sociodemographic risk factors. The authors reported that exposure to severe intimate partner violence as an adolescent significantly increased the odds of alcohol-use problems in early adulthood for young women (odds ratio = 5.63) but not for young men. Exposure to intimate partner violence did not increase the odds of other substance-use indicators for either gender. The authors concluded that girls exposed to intimate partner violence may be at increased risk for problems with alcohol use in adulthood and should be a target for prevention and intervention efforts. Overall, however, the association between exposure to intimate partner violence and later substance-use problems is less than anticipated in this high-risk community sample [58].

Intimate partner violence is also associated with a higher risk of substance use by adults. However, it is less clear how the use of specific substances relates to relationship violence. Smith et al. analyzed data from wave two of the National Epidemiologic Survey on Alcohol and Related Conditions (2004−05) and tested associations between substance use disorders and intimate partner violence using logistic regression models while controlling for important covariates and accounting for the complex survey design. Alcohol use disorders and cocaine use disorders were most strongly associated with intimate partner violence perpetration, while cannabis use disorders and opioid use disorders were most strongly associated with intimate partner violence

victimization. Overall, substance use disorders were consistently related to intimate partner violence after controlling for important covariables [59].

Peer Pressure and Addiction

Studies have shown that parental supervision and monitoring resulted in lower alcohol or substance abuse among 8th and 10th graders. However, one developmental transition characteristic of adolescence is moving from home after finishing high school and their increasing involvement with peers. Studies have indicated that peer use of alcohol and or drug is a strong predictor of alcohol and or substance abuse by young adults. In one study, the authors concluded that predictors of drinking and heavy episodic drinking during the first year after high school included being white, living on campus, previous drinking history, lower parental expectation, and having peers who drink [60]. Guo et al. reported that on average randomly assigned to a drinking peer as opposed to no drinking peer increased college binge drinking by 20%−40% per month. However, such effect was observed among youths with a medium level of a genetic propensity for alcohol use indicating that gene-environment analyses can uncover social-contextual effects likely to be missed by the traditional sociological approach [61].

Because alcohol use typically is initiated during adolescence as well as young adulthood and may have long-term consequences, the Monitoring the Future (MTF) study annually assesses various measures of alcohol use among 8th-, 10th-, and 12th-grade students. These analyses have found that although alcohol use among these age groups overall has been declining since 1975, levels remain high. Thus, in 2011 about one-quarter of 8th graders, one-half of 10th graders, and almost two-thirds of 12th graders reported drinking alcohol in the month preceding the interview. Binge drinking (i.e., consumption of five or more drinks in one occasion by a male or four and more drinks by a female in a single occasion) was also prevalent. The MTF study has also identified numerous factors that influence the risk of alcohol use among adolescents, including parents and peers, school and work, religiosity and community attachment, exercise and sports participation, externalizing behavior and other drug use, risk taking and sensation seeking, well-being, drinking attitudes and reasons for alcohol use. Drinking during adolescence can have long-term effects on a person's life trajectory. Moreover, misbehavior and peer encouragement of misbehavior were positively associated with substance abuse. Moreover, school disengagement, failure in school or skipping school also increase the risk of substance abuse. In contrast, good grades, education expectations, and school bonding protect from alcohol or substance abuse. Studies have shown that religiosity protects adolescence from alcohol and substance abuse. Good self-esteem also protects adolescence from alcohol or substance abuse [62].

All adolescents who participate in bullying whether they are the perpetrators, victim, or a combination of both roles have been shown to have a higher risk of psychological problems later in life compared to those who do not participate in bullying or are victims of bullying. Research has shown that females are more likely to be bulled via verbal attacks whereas males are physically bullied. Studies indicated that playing the role of bully has been positively associated with increased alcohol use. Stone and Carlisle based on a review of a data of 7585 subjects observed that racial bully perpetrators were most likely to have used cigarettes, alcohol, and marijuana followed by youths in mixed victim/perpetrator groups [63]. Gang affiliation is also associated with a higher risk of alcohol and substance use. Specifically, higher rates of alcohol and marijuana have been reported among gang members than among those affiliated with a group of deviant peers. Alcohol advertisement is also associated with higher alcohol use by young adults. Ross et al. concluded that there is a robust relationship between underage youth's brand-specific exposure to alcohol advertisement on television and their consumption of those same alcohol brands during the past 30 days [64].

Waylen et al. investigated whether exposure to alcohol use in films (AUFs) is associated with early alcohol use, binge drinking, and alcohol-related problems in British adolescents. The authors utilized a cross-sectional study with 5163 subjects who were 15-year-olds from the Avon Longitudinal Study of Parents and Children in the United Kingdom. The authors reported that after adjustment, adolescents with the highest exposure to AUFs were 1.2 times more likely to have tried alcohol and 1.7 times more likely to engage in binge drinking compared with those least exposed. In addition, adolescents with the highest exposure to AUFs were 2.4 times more likely to drink weekly and 2.0 times more likely to have alcohol-related problems than those least exposed [65]. Various risk factors for adolescent alcohol or drug abuse are summarized in Table 5.2.

However, there are environmental factors that may protect an adolescent from alcohol or drug abuse. Good performance in school, good communication with parents, coming from an intact family, parental supervision, and participation of adolescents in religious activities all have protective effects. Many studies have shown that higher levels of religiosity have been associated with lower rates of alcohol abuse, marijuana abuse, and cigarette smoking among adolescents. Rostosky et al. also observed the protective effect of religiosity on alcohol and substance abuse in adolescents but such protection was only observed in heterosexual adolescents [66].

Yoga/Meditation Protects Adolescents and Adults From Substance Abuse

The prevalence of substance use among adolescents remains high. Several biological, physical, and social-contextual factors interact to make adolescence a particularly vulnerable time period for substance use initiation, such as the onset of puberty, increases in unsupervised time with peers, and a reorganization of family relationships [67]. The prefrontal neural circuitry, which is responsible for exerting effortful control over behavior in emotional contexts, is not fully developed in adolescence and as a result, increased sensitivity to reward and decreased ability to inhibit responses may increase the vulnerability of an adolescent to experiment with illicit drugs for seeking pleasure [68]. A combination of these neurobiological changes with social-contextual risk factors such as low parental monitoring and/or deviant peer environments, as well as psychological risk factors such as high anxiety or mood impairment may result in increased substance use during adolescence [67].

To address these concerns, school-based programs designed to prevent substance use have grown substantially in the past 20 years. The majority of school-based prevention programs employs behavioral skills training by providing students with information about substance use, and may also include modeling of skills by the teacher or a guest speaker, student experiential practice of skills within the classroom, discussions about skill performance, and/or practice in settings outside of the classroom (i.e., homework). However, such trainings are not sufficient to prevent substance abuse in adolescents. Review papers published in the medical literature have suggested that mind—body practices such as yoga and meditation could be beneficial for preventing and treating substance abuse and addictive behavior. The mechanism of the protective effect of yoga and meditation may be due to multiple factors including stress reduction and mood improvement via modulating neuroendocrine system as well as improving self-awareness and self-control [69,70]. As a result, a number of substance abuse treatment programs have incorporated yoga or meditation as

TABLE 5.2
Various Risk Factors for Adolescent Alcohol or Drug Abuse.

Risk Factor	Comments
Maternal abuse of alcohol	Fetal alcohol syndrome or fetal alcohol spectrum of disorders may result due to maternal alcohol use. Offspring of mothers who drank during pregnancy have a higher risk of alcohol or substance abuse during adolescent years. Moreover, such offspring may experience pleasure from even alcoholic odor.
Maternal opioid/drug abuse	Maternal opioid abuse may cause stillbirth or birth defects in offspring. Moreover, maternal opioid/drug abuse may increase the susceptibility of offspring to abuse drugs during adolescent years or adulthood.
Alcoholic parents or drug abusers	Alcohol or substance use by parents increases the risk of alcohol/substance abuse in offspring during adolescent years.
Single mother/single-parent family	Young adults living with one parent have a higher risk of substance abuse. However, strict parental rules deter adolescents from alcohol use.
Childhood supervision neglect	May increase the risk of alcohol or drug abuse during adolescent years.
Childhood emotional/physical abuse	May increase the risk of alcohol or drug abuse during adolescence years
Sexual abuse	Although less common than other childhood maltreatment, sexual abuse significantly increases the risk of drug or alcohol abuse during adolescent years.
Intimate partner violence	Exposure of adolescents to parental conflict/violence is associated with an anxiety disorder in males but alcohol and substance abuse by females indicating that females are more affected than males.
Peer pressure for alcohol/substance abuse	Sometimes peer pressure or desire to be popular may increase the risk of substance abuse especially if an individual believes that substance abuse may make him more popular with the group. In contrast, good grades and living with a roommate who does not drink or abuse drugs protects an individual from drinking or substance abuse.
Bullying	Males are involved in bullying more than females. Playing the role of a bully is associated with a higher risk of alcohol abuse.
Gang affiliation	Gang affiliation results in a significantly increased risk of substance abuse.
Availability of alcohol	If alcohol is easily available at home because parents are regular drinkers it may increase the risk of alcohol use by adolescents. In contrast, if parents are teetotalers, it prevents adolescents from drinking.
Exposure to alcohol use in films	A British study indicates that adolescents who watch films where people are drinking increases the risk of alcohol use by these adolescents.

contributing therapies, and there is some research demonstrating the efficacy of such treatments for adults [71]. In addition, a number of studies have evaluated the specific effects of yoga interventions on addictive behavior, substance use risk factors, and substance abuse itself and have found positive results for adults, such as improvements in quality of life in opioid-dependent users.

Dhawan et al. investigated changes in quality of life in treatment-seeking male opioid-dependent users ($n = 55$; study group) following the practice of a yogic breathing exercise program by comparing results with a control group ($n = 29$). The study group besides standard treatment (long-term pharmacotherapy with buprenorphine in flexible dosing schedule) was taught yoga breathing for 3 days while the control group received standard treatment alone. Assessments were made at baseline and at 3 and 6 months. The authors observed significant improvement in the quality of life in subjects who practiced yoga compared to the control group. In addition, urine screening results were negative for the study group indicating no drug use at 6 months. The authors concluded that yoga practice is useful in drug rehabilitation program [72].

Butzer et al. tested the efficacy of yoga for reducing substance use risk factors during early adolescence. Seventh-grade students in a public school were randomly assigned to receive either a 32-session yoga intervention ($n = 117$) in place of their regular physical education classes or to continue with physical-education-as-usual ($n = 94$; control group). Participants (63.2% female; 53.6% White) completed pre- and postintervention questionnaires assessing emotional self-regulation, perceived stress, mood impairment, impulsivity, desire for substance abuse, and actual substance use. Participants also completed questionnaires at 6-month and 1-year postintervention. Results revealed that participants in the control condition were significantly more willing to try smoking cigarettes immediately postintervention than participants in the yoga condition. Interestingly, long-term follow-up analyses revealed a pattern of delayed effects in which females in the yoga group, and males in the control group, demonstrated improvements in emotional self-control. The findings suggest that school-based yoga may have beneficial effects regarding preventing males' and females' willingness to smoke cigarettes, as well as improving emotional self-control in females [73].

Emotional regulation involves self-awareness, cognitive reframing, and mindfulness, all of which increase one's capacity to control, override, or accept spontaneous negative emotional responses. Yoga has been suggested to be an exceptional path toward skillful emotional regulation and thus may provide a highly effective route toward alleviating alcohol or drug misuse. For example, a randomized controlled trial found that alcohol-dependent participants in an 8-week yoga intervention had greater declines in dependence severity compared to participants in the physical training exercise control condition [74]. Individuals with PTSD often exhibit high-risk substance use behaviors. In one study, the authors investigated the effect of a yoga intervention on alcohol and drug abuse behaviors in women with PTSD by analyzing data from a pilot randomized controlled trial comparing a 12-session yoga intervention with an assessment control for women age 18−65 years with PTSD. The authors concluded that results from the pilot study suggest that specialized yoga therapy may play a role in attenuating the symptoms of PTSD, reducing the risk of alcohol and drug use, and promoting interest in evidence-based psychotherapy [75]. Symptoms related to impaired self-control involve reduced activity in the anterior cingulate cortex (ACC), adjacent prefrontal cortex (mPFC) and other brain areas. Behavioral training such as mindfulness meditation can increase the

function of control networks including those leading to improved emotion regulation and thus may be a promising approach for the treatment of addiction. Emerging evidence has shown that mindfulness meditation induces increased connectivity and activity in ACC/mPFC regions of the brain thus promoting emotion regulation and improving self-control related brain activity that can help in the prevention and treatment of addictions such as tobacco, alcohol, cocaine, as well as other various behavioral disorders including obesity, gambling, and excessive use of the internet that are also associated with self-control deficits [76].

STRESS INCREASES RISK OF SUBSTANCE ABUSE

The body responds to stress with self-regulating processes that contain both physiological and behavioral components. In response to stress, neurons in the paraventricular nucleus release two hormones: corticotropin-releasing factor and arginine vasopressin into the blood vessels connecting to the hypothalamus and pituitary. Both hormones stimulate anterior pituitary and as a result, ACTH is secreted into the blood that eventually results in the secretion of adrenocortical hormones mainly cortisol in humans. In summary, the stress hormone cortisol is produced due to the activation of the HPA-axis.

Cortisol can bind with two types of receptors: Type I receptors (mineralocorticoid receptors with high affinity for cortisol) and Type-II (glucocorticoid receptors with low affinity for cortisol). During normal activity cortisol usually binds with Type I receptor thus maintaining the required basal level of cortisol needed for daily activity but when cortisol level is high in response to stress, it binds with Type II receptors and such binding also terminates the stress signal (negative feedback) that is essential to maintain blood cortisol to a predetermined set point. A healthy acute stress response is characterized by a quick rise in blood cortisol level followed by a decline of cortisol level to the basal level with the termination of stressful events. However, when a person is exposed to chronic stress, it may deregulate the normal activity of the HPA-axis that may result in metabolic and/or neuropsychiatric problem including susceptibility to alcohol or drug abuse.

Three factors including the genetic makeup of a person, early life environment and current life stress determine the function of the HPA-axis. Maternal use of alcohol may impair HPA-axis responsivity in adulthood. Maternal stress during gestation may also modify HPA-axis responsivity during childhood as well as adulthood. Childhood traumas including physical abuse or sexual abuse adversely affect the developing brain and may permanently alter the normal stress response capacity of the HPA-axis. This may explain why childhood adverse events may increase the risk of alcohol and drug abuse in adulthood. Independent of prenatal and childhood stressors, periods of severe stress in an adult such as family and work-related problems, neighborhood violence, combat exposure, and chronic illness may also alter HPA-axis dynamics and may increase the cortisol burden of the body. This alteration in the function of the HPA-axis causes higher cortisol exposure or greater cortisol burden following each stressful event [77].

Extensive research in both clinical and preclinical demonstrated that stress is associated with increased risk of alcohol and drug abuse. Animal experiments have shown acquisition to amphetamine and cocaine were enhanced when rats were subjected to stress such as social isolation or tail pinch or born to female rats restrained during pregnancy or male rats exposed to attack by an aggressive male rat. Therefore, activation of the HPA-axis by stress may be linked to susceptibility to substance abuse. One explanation for the high concordance between stress-related disorders and drug addiction is the self-medication hypothesis, which suggests that a person may use alcohol or drug to cope with the tension associated with life stressor or to relieve symptoms of anxiety and depression resulting from a traumatic life event. Another characteristic of self-administration of drug or alcohol is that the person has direct control on how much substance to abuse and its subsequent effect on the HPA-axis activation. This controlled activation of the HPA-axis may result in the production of an internal state of arousal or stimulation that is intended by that individual (sensation seeking hypothesis). During abstinence exposure to a stressor again or drug/alcohol-associated cue may stimulate the HPA-axis to remind the individual regarding the pleasurable effect of drug or alcohol in the past that eventually may lead to craving or relapse [78].

Stressful life events are important risk factors for the development of neuropsychiatric disorders that include affective disorders and addiction to food and illicit substances. Such events are associated with significant changes in cognitive, neurotransmitter, and neuroendocrine systems in humans. Life stresses are not only risk factors for the development of addiction but also are triggers for relapse to drug use. The development and clinical course of addiction-related disorders do appear to involve neuroadaptations within neurocircuitry that modulates stress responses and are influenced by several neuropeptides. These include corticotropin-releasing factor, the prototypic member of this class, as well as oxytocin and arginine-vasopressin that play important roles in affiliative behaviors. Interestingly, these peptides function to balance emotional behavior, with sexual dimorphism in the oxytocin/arginine-vasopressin systems, a pathway that might play an important role in the differential responses of women and men to stressful stimuli and the specific sex-based prevalence of certain addictive disorders. Carroll et al. reported that relapse to smoking, alcohol, and cocaine use was more strongly linked to stress and depression in women than in men because females seem to be more sensitive to the rewarding effects of drugs than males, and estrogen is a major factor that may explain these sex differences [79].

In one study, the authors investigated the effects of distinct forms of stress on nonmedical use of prescription drugs by collecting data from 5308 young adult men from the Swiss cohort study on substance use risk factors. Various forms of stress (discrete, potentially traumatic events, recent, and long-lasting social-environmental stressors) during the period preceding the nonmedical opioid use were measured. The authors observed that nonmedical use of prescription drugs was significantly associated with the cumulative number of potentially traumatic events, with problems within the family, and the peer group. Sexual assault by acquaintances was associated only with abuse of tranquilizers. The authors concluded that nonmedical use of prescription drugs appears to be more consistently associated with discrete and potentially traumatic events and with recent social-environmental stressors than with long-lasting stressors due to family malfunctioning during childhood and youth. Physical and sexual assaults perpetrated by strangers showed more associations with nonmedical use of prescription drugs than those perpetrated by a family member [80].

Smith et al. analyzed the data from wave 2 of the National Epidemiologic Survey on Alcohol and Related Conditions, a nationally representative sample of noninstitutionalized adults, to examine the relationship between PTSD diagnosis and nonmedical opioid use. The authors reported that in the adjusted model, a past year PTSD diagnosis was associated with higher odds of past year nonmedical opioid use for women and men, but the association was stronger for women. In addition, PTSD was associated with higher odds of an opioid use disorder diagnosis for women, but not for men. Regarding the relationship between specific symptom clusters among those with a past year PTSD diagnosis, the important sex difference was observed.

For women, the avoidance symptom cluster was associated with higher odds of nonmedical opioid use, an opioid use disorder diagnosis, and a higher rate of average monthly frequency of nonmedical opioid use, while for men the arousal/reactivity cluster was associated with higher odds of nonmedical opioid use, an opioid use disorder diagnosis, and a higher rate of average monthly frequency of nonmedical opioid use. In addition, for men, the avoidance symptom cluster was associated with higher odds of an opioid use disorder diagnosis, but a lower rate of average monthly frequency of nonmedical opioid use. The authors concluded that PTSD is more strongly associated with substance use for women than men. Further, results based on individual symptom clusters suggest that men and women with PTSD may be motivated to use substances for different reasons [81].

Terrorist attacks may also increase alcohol use among people. Studies have shown that following the terrorist attack that destroyed the World Trade Center in New York in 2001, alcohol consumption was increased in New York City and elsewhere for a short term after the attack. Longer-term studies showed increased alcohol consumption 1 and 2 years later among New Yorkers who had greater exposure to the attack [82]. Grieger et al. examined posttraumatic stress PTSD, alcohol use, and perceptions of safety in a sample of survivors of September 11, 2001, the terrorist attack on the Pentagon by surveying 77 survivors 7 months after the attack. The authors reported that 11 respondents (14%) had PTSD. Those with PTSD reported higher levels of initial emotional response and peritraumatic dissociation. Ten respondents (13%) reported increased use of alcohol. Women were more than five times as likely as men to have PTSD and almost seven times as likely to report increased use of alcohol. Persons with higher peritraumatic dissociation were more likely to develop PTSD and report increased alcohol use. Those with lower perceived safety at 7 months had a higher initial emotional response and greater peritraumatic dissociation and were more likely to have PTSD, to have increased alcohol use, and to be female [83].

Stressful Life Events and Alcohol or Substance Abuse

Stressful life events may also increase the risk of substance abuse. In addition, substance abuse and alcohol use are associated with domestic partner abuse, violence, and crime. Many investigators reported a close link between violent behavior, homicide, and alcohol intoxication. Studies conducted on convicted murders suggest that about half of them were under the heavy influence of alcohol at the time of murder [84]. Various life stressors that may increase susceptibility to alcohol or drug abuse are listed in Table 5.3.

Problematic substance use increases the likelihood of unemployment and decreases the chance of finding and holding down a job. Moreover, unemployment is a significant risk factor for substance use and the subsequent development of substance use disorders [85]. Compton et al. based on data analysis of 405,000 noninstitutionalized adult participants in the 2002–10 U.S. National Survey on Drug Use and Health, reported higher rates of past month tobacco and illicit drug use, heavy alcohol use, and past-year drug or alcohol abuse/dependence among the unemployed. The authors concluded that employment status was strongly and robustly associated with problematic use of substances [86].

Bruguera et al. conducted a questionnaire-based survey in drug dependence treatment settings, in three geographically different jurisdictions (England, Catalonia, and Poland), using 180 drug users. The authors observed that most of the participants of the survey (58.3%) reported an increase in drug use during the crisis as a result of economic recession, compared with only 25.6% of the sample who reported a decrease in drug use. Those who reported cutting down on the amount of drug use during the economic recession reported economic difficulties as the main reason, but illegal drug use reduction was compensated by increased smoking in 46.3% of the patients and increased alcohol use in 39.4%. Many factors contributed to increased alcohol and substance abuse during the economic recession including fear of losing a job as well as family and friends' economic problems [87]. Living alone is associated with a substantially increased risk of alcohol-related mortality, irrespective of gender, socioeconomic status, or the specific cause of death [88].

Domestic violence is the use of intentional emotional, psychological, sexual, or physical force by one family member or intimate partner to control another. Violent acts include verbal, emotional, and physical intimidation; destruction of the victim's possessions; threats; forced sex; and slapping, punching, kicking, choking, burning, stabbing, shooting, and killing victims. Spouses, parents, stepparents, children, siblings, elderly relatives, and intimate partners may all be targets of domestic violence. One of the major causes of domestic violence is alcohol abuse. However, it is not clear whether a man is violent because he is drunk or whether he drinks to reduce his inhibitions against his violent behavior [89].

TABLE 5.3
Life Stressors That May Increase Risk of AAlcohol or Substance Abuse.

Stressful Life Event	Comment	Reference
Unemployment	Substance use increases the likelihood of unemployment and, at the same time, prolonged unemployment increases the risk of substance abuse.	[84,85]
Economic recession/ financial trouble	An economic recession or financial trouble is associated with significant stress that may increase susceptibility for alcohol or drug abuse.	[86]
Living alone/ social isolation	Increased risk of alcohol abuse and increased risk of mortality from alcohol abuse.	[87]
Domestic violence	Alcohol and drug abuse are associated with increased risk of domestic violence. Women who are subjected to intimate partner violence are more likely to abuse drugs during pregnancy.	[88–90]
Divorce/ widowhood	Divorce significantly increases the risk of alcohol and drug abuse. Widowhood also to a lesser extent increases the risk of alcohol and drug abuse.	[91,92]
High crime neighborhood	Living in a high crime neighborhood increases the risk of alcohol and drug abuse especially among young adults.	[93,94]

Research indicates links between women's experiences of intimate partner violence and their use of drugs. In one study, the authors investigated that 85 prenatal care patients to describe the women's use of alcohol and illicit drugs, both before and during pregnancy, about their experiences of various types of intimate partner violence before and during pregnancy (including psychological aggression, physical abuse, and sexual coercion). The Conflict Tactics Scales 2 was used to assess the women's experiences of intimate partner violence. The women were asked about their frequency of alcohol use, and alcohol using women were administered a short version of the Michigan Alcohol Screening Test to assess the women for symptoms of alcohol disorder. The women's use of illicit drugs was assessed by asking the women about their frequencies of various types of drug use, and drug-using women were administered the Drug Abuse Screening Test to assess the women for symptoms of drug use disorder. The results showed that before pregnancy, women who were physically assaulted by their partners were somewhat more likely to drink alcohol and use illicit drugs compared with women who did not experience such violence. After the women became pregnant, the links between women's experiences of intimate partner violence and their use of substances became stronger, with the women who experienced partner violence being more likely to use both alcohol and illicit drugs. Furthermore, among the substance-using women, those who were psychologically and physically abused had somewhat elevated levels of substance disorder symptoms during pregnancy compared with women who did not suffer such victimization [90].

Drug use has been shown to interact in complex ways with the occurrence and prevalence of family and domestic violence, with illicit drug use being associated with an increased risk for domestic violence. In an Australian study, the authors recruited 5118 subjects through an online survey panel who completed an online self-report survey assessing the role of alcohol and other drugs on violence, with a specific focus on domestic violence. The authors using binary logistic regression showed that respondents who reported having used any illicit drug in the past 12 months (with or without alcohol use) had over three times the odds of experiencing any violence in the past 12 months (odds ratio: 3.18) compared with those not using illicit drugs. For the most recent domestic violence incident, age group was the only significant demographic predictor of drug involvement at this incident; younger age groups were over twice more likely to report drug involvement than those over 65 years old. The authors concluded that drug use increases both the risk and impact of domestic violence [91].

Divorce increases the risk of both alcohol and substance abuse. Kendler et al. reported that divorce was strongly associated with risk for first alcohol use disorder onset in both men (hazard ratio: 5.98) and women (hazard ratio: 7.29). In addition, divorce was also associated with the recurrence of alcohol use disorder in those who suffered from alcohol use disorder before marriage. Moreover, widowhood increased risk for alcohol use disorder in both men (hazard ratio: 3.85) and women (hazard ratio: 4.10). Among divorced individuals, remarriage was associated with a large decline in alcohol use disorder in both sexes (men: hazard ratio: 0.56, women: hazard ratio: 0.61) [92].

Rates of drug abuse are higher among divorced individuals than among those who are married, but it is not clear whether divorce itself is a risk factor for drug abuse or whether the observed association is confounded by other factors. Edwards et al. examined the association between divorce and onset of drug abuse in a population-based Swedish cohort born during 1965–75 ($n = 651,092$) using Cox proportional hazards methods, with marital status as a time-varying covariate. Potential confounders (e.g., demographics, adolescent deviance, and family history of drug abuse) were included as covariates. Parallel analyses were conducted for widowhood and drug-abuse onset. In models with adjustments, the authors observed that divorce was associated with a substantial increase in the risk of drug-abuse onset in both sexes (hazard ratios > 5). Widowhood also increased the risk of drug-abuse onset, although to a lesser extent [93].

African American youth living in urban environments with exposure to drug activity, violence, and neighborhood disorder are at increased risk for both initiation and progression to more frequent and problematic marijuana use during high school [94]. Living in disorganized neighborhoods characterized by high levels of poverty, crime, violence, and deteriorating buildings has been associated with increased alcohol consumption and mental health problems. Cambron et al. analyzed data drawn from the Seattle Social Development Project ($n = 790$), a theory-driven longitudinal study originating in Seattle, WA, to estimate trajectories of AUD symptoms in subjects from age 21 to 39. The authors reported that on average, AUD symptoms decreased as individuals got older. Living in more disorganized neighborhoods and experiencing psychological distress was associated with increased AUD symptoms. Results of mediation analysis suggested that

psychological distress is a mechanism by which disorganized neighborhoods increased the risk of AUD from age 21 to 39 [95].

Happy Marriage, Religiosity, and Spirituality Protects People from Substance Abuse

A happy marriage has a protective effect on alcohol or substance abuse. Kendler et al. based on a population study involving 3,220,628 individuals concluded that first marriage to a spouse with no lifetime alcohol use disorder is associated with a large reduction in risk of alcohol use disorder. These observations are consistent with the hypothesis that the psychological and social aspects of marriage and in particular health-monitoring spousal interactions, strongly protect against the development of alcohol use disorder [96]. In another study, the authors conducted multinomial logistic regression on data from a national sample of 34,650 adults mostly between the ages of 18 and 35, collected through the 2007 National Survey on Drug Use and Health. The authors reported that white males who were lesser educated and living in poverty were more likely to exhibit co-occurring substance abuse and psychological distress than their demographically similar counterparts. However, being married protect from potential substance abuse [97].

Hearld et al. based on the analysis of data collected from The Collaborative Psychiatric Epidemiology Surveys, sponsored by the National Institute of Mental Health, a repository of race, ethnicity, and mental health data, reported that racial and ethnic minorities were found to have lower rates of substance use and abuse compared to Whites, and foreign-born individuals were consistently less likely to use or abuse substances compared to American-born minorities. Mental health conditions were highly associated with substance use and abuse, and social support by way of religious participation and marriage was protective against substance use and abuse [98].

Adolescents are exposed to various stressors that may increase the risk of substance use. Debnam et al. explored the association between substance use and stress among male and female high school students in relation to spirituality as a moderator. The authors used cross-sectional data collected from 27,874 high school students (male: 50.7%, female: 49.3%) across 58 high schools in Maryland that included an ethnically diverse sample (49% Caucasian, 30% African American) with an average age of 16 years old. Higher rates of substance use were generally found among male students compared to female students. Multilevel analyses indicated a positive association between stress and substance use among male and female students after adjusting for demographic and school-level factors. Both male and female students who reported turning to spiritual beliefs when experiencing substance abuse problems were less likely to use illicit substances. However, the interaction between stress and spirituality was significant for males only. The authors concluded that spirituality might be a viable coping mechanism useful for helping high school students adapt to stressful circumstances and avoid substance abuse [99]. Spirituality is generally protective against the initiation of alcohol and drug use and progression to disordered use. In addition, mutual-help organizations, such as Alcoholics Anonymous, were founded on spiritual principles, and reliance on a "higher power" is a central component of the 12 steps to achieve sober state [100].

PERSONALITY TYPE, MENTAL DISORDER AND RISK FOR SUBSTANCE ABUSE

The Five-Factor Model of personality developed by McCrae and John is one of the most commonly used models in psychology. The model consists of five personality factors that can be evaluated using the revised NEO Personality Inventory Facet Scale [101].

- Openness
- Conscientiousness
- Extraversion
- Agreeableness
- Neuroticism

Gore and Widiger commented that DSM-5 (fifth edition published by the American Psychiatric Association) maladaptive trait dimensional model proposal that included 25 traits organized within five broad domains (i.e., negative affectivity, detachment, antagonism, disinhibition, and psychoticism) may align with the five-factor model. For example, negative affectivity would align with the neuroticism of the five-factor model while disinhibition would align with low conscientiousness. The authors based on a study of 445 undergraduates along with the personality inventory of DSM-5 concluded that the results provided support for the hypothesis that all five domains of the DSM-5 dimensional trait model are maladaptive variants of five-factor model general personality structure [102]. A substantial body of research also indicates that the personality disorders included within the DSM-5 can be understood as extreme and/or maladaptive variants of the five-factor model [103].

Martin and Sher based on a study of 468 young adults observed that alcohol use disorder was positively associated with neuroticism and negatively associated with agreeableness and conscientiousness [104]. Terracciano et al. observed that compared to never users, current marijuana abusers scored higher on openness, average on neuroticism, and lower on agreeableness as well as conscientiousness. Compared to never users, current cocaine users scored higher on neuroticism and lower on conscientiousness. The authors concluded that high levels of negative affect and impulsive traits were associated with substance abuse. Moreover, there was also a link between low scores on conscientiousness and substance use [105]. Users of drugs are more prone to negative emotions (neuroticism) and tend to be distrustful, manipulative, unreliable and undisciplined. Sutin et al. also observed the same pattern of elevated neuroticism, decreased agreeableness, and decreased conscientiousness among current users of cocaine/heroin [106].

Impulsivity is related to risk-taking, quick decision making, and lack of planning. Impulsivity is closely linked to alcohol and substance abuse, both as a contributor to use and as consequences of use. Bozkurt et al. based on a study of alcohol-dependent patients ($n = 94$) and healthy controls ($n = 63$), observed that the mean impulsivity score was higher in alcohol-dependent subjects than control (BIS-11 score; 69.34 in alcohol-dependent subjects vs. 58.81 in health controls) [107].

One personality trait that stands out as a contributing factor to increased risk of substance abuse is novelty seeking. Novelty seeking, affected by both genetic and environmental factors, is defined as the tendency to desire novel stimuli and environments. It can be measured in humans through questionnaires. On the molecular level, both novelty seeking and addiction are modulated by the central reward system in the brain. Dopamine is the primary neurotransmitter involved in the overlapping neural substrates of both parameters. Therefore, the novelty-seeking trait can be used for predicting individual vulnerability to drug addiction and for generating successful treatment for patients with substance abuse disorders [108].

Liang et al. commented that novelty seeking is a core personality trait that primes the susceptibility to drug addiction. Evidence supports the association between dopamine and novelty seeking. Opioid-dependent patients show higher levels of novelty seeking, and repeated opioid exposure can cause cognitive deficits including poor cognitive flexibility and impaired impulse control. The authors recruited 22 opioid-dependent individuals and 30 age- and sex-matched healthy controls for the study. Single-photon emission computed tomography with ^{99m}Tc-TRODAT-1as a ligand was used to measure the striatal dopamine transporter availability. The trail making test (TMT) was performed to assess cognitive flexibility. Cloninger's Tridimensional Personality Questionnaire was used to measure novelty seeking. The authors observed that in opioid-dependent patients, the striatal dopamine transporter availability was lower and negatively associated with the TMT score. Moreover, an inverted-U shape graph significantly matched the scores of novelty seeking as a function of the striatal dopamine transporter availability, with maximum novelty seeking potential in the midrange of the dopamine transporter availability. An extra sum-of-squares F test was conducted, indicating that a quadratic model fitted the association between the dopamine transporter availability and novelty seeking better than a linear model did. In brief, in opioid-dependent patients, the striatal dopamine transporter availability was nonlinearly linked to novelty seeking and linearly linked to cognitive flexibility [109].

Comorbid Substance Use Disorder and Personality Disorder

When two disorders occur simultaneously in the same person, it is referred to as comorbid. Many individuals with substance use disorder have comorbid mental illness and vice versa. There are several personality disorders and psychiatric disorders that are associated with a higher prevalence of substance use disorders. Smith et al also observed a higher prevalence of substance abuse among people with antisocial personality disorder (ASPD). This type of personality disorder is characterized by a pervasive pattern of disregard for and violation of rights of others occurring since age 15 in persons at least 18 years old. In general, fulfilling three of the seven criteria (failure to follow social norms, deceitfulness, impulsivity, irritability, and aggressiveness, reckless disregard for the safety of the person and others, consistent irresponsibility and lack of remorse after illicit behavior) can establish that the person has ASPD. It has been estimated that 1%–3% of the general population (3.8%–6.8% among males and 0.8%–1% in females) may have ASPD but in the prison population the prevalence is estimated to be 35%–47% and prevalence is 18%–40% among substance-dependent individuals. In one study involving the rural population using nonmedical prescription opioids, the authors observed that 31% had ASPD. In multivariant analyses, the authors observed that distrust and conflict within an individual's social network, as well as past 30 days use of heroin and crack cocaine, male gender, younger age, lesser education, heterosexual orientation, and comorbid major depressive disorder were associated with meeting diagnostic criteria of ASPD [110].

A borderline personality disorder is characterized by a pervasive pattern of instability in affect regulation, impulse control, interpersonal relationships, and self-image. Clinical signs of the disorder include emotional dysregulation, impulsive aggression, repeated self-injury, and chronic suicidal tendencies. Borderline personality disorder (BPD) affects 2.7% of adults. The most frequent comorbid psychiatric disorders in BPD patients are anxiety and affective disorders, including posttraumatic stress disorder. About 78% of adults with BPD also develop a substance-related disorder or addiction at some time in their lives. These persons are more impulsive and clinically less stable than BPD patients without substance dependency [111].

Decades of research has shown that anxiety disorders and substance abuse disorders co-occur at greater rates than expected by chance alone. Anxiety disorders are more strongly associated with substance dependence than substance abuse. Generalized anxiety disorders and panic disorders with or without agoraphobia showed the highest association with substance use disorders. There is also an association between obsessive-compulsive disorder and alcohol use [112]. Patients suffering from bipolar disorder are also at higher risk of substance abuse. In general, the male gender, history of a higher number of manic episodes, and suicidality are associated with higher susceptibility to substance use disorder [113].

The presentation of major depressive disorder is often complicated by the co-occurrence of substance abuse disorder such as alcohol or illicit drug abuse or dependence. Nearly one-third of the patients with major depressive disorder also have substance abuse disorder and the comorbidity is associated with a higher rate of suicide and greater social or personal impairment [114]. Dysthymic disorder is defined as a low-grade chronic depression that lasts for 2 years for adults. This disorder has a prevalence of 6% in the general population. There is an elevated comorbidity rate between various mood disorders including dysthymia and substance abuse. Cassidy et al. concluded that substance abuse is a major comorbidity in bipolar patients with nearly 60% reporting a history of some lifetime substance abuse [115].

Toftdahl et al. based on a survey of 463,003 psychiatric patients in Danish Register observed that prevalence of any lifetime substance abuse disorder was 37% for schizophrenia, 35% schizotypal disorder, 28% for other psychoses, 32% for bipolar disorder, 25% for depression, 25% for anxiety, 11% for the obsessive-compulsive disorder, 17% for PTSD, and 46% for personality disorder. These personality disorders included paranoid, schizoid, dissocial, emotionally unstable, histrionic anankastic, anxious, dependent, unspecified, and mixed disorders [116]. Mauri et al. also reported that 34.7% of the first episode schizophrenia patients had a lifetime history of substance abuse. The age of onset of schizophrenia was significantly lower for drug abusers than patients without any type of abuse and for alcohol abusers. In multidrug users, cannabis was used more frequently (49%) followed by alcohol (13%) and cocaine (4%) [117].

Substance abuse also increases the risk of schizophrenia. Nielsen et al. investigated whether substance abuse increases the risk of developing schizophrenia. The longitudinal, nationwide Danish registers were linked to establish a cohort of 3,133,968 individuals, identifying 204,505 individuals diagnosed with substance abuse and 21,305 diagnosed with schizophrenia. Data analysis by the authors showed that a diagnosis of substance abuse increased the overall risk of developing schizophrenia with a hazard ratio of 6.04. Although cannabis and alcohol abuse showed the strongest association, abuse of hallucinogens, sedatives, and other illicit substances also increased the risk significantly [118].

Substance use disorders occur commonly in patients with schizophrenia and dramatically worsen their overall clinical course. Although the exact mechanisms contributing to substance use in schizophrenia are not known, several theories have been put forward to explain the basis of the co-occurrence of these disorders. Khokhar et al. proposed a unifying hypothesis that combines recent evidence from epidemiological and genetic association studies with brain imaging and preclinical studies to provide an updated formulation regarding the basis of substance use in patients with schizophrenia. The authors suggested that the genetic determinants of risk for schizophrenia (especially within neural systems that contribute to the risk for both psychosis and addiction) make patients vulnerable to substance use. As this vulnerability may arise before the appearance of psychotic symptoms, increased use of illicit drugs in adolescence may both enhance the risk for developing a later substance use disorder and also serve as an additional risk factor for the appearance of psychotic symptoms [119].

CONCLUSIONS

Alcohol and drug abuse including opioid abuse is a serious public health issue. Many environmental factors increase the risk of substance abuse. Maternal use of opioids is not only associated with potential birth defects but also neonatal withdrawal symptoms requiring extended hospitalization and treatment with buprenorphine, methadone, or other medications. Childhood maltreatment including physical neglect, emotional, and physical abuse, as well as sexual abuse, significantly increase the risk of substance abuse later in life. In addition, various stressors of life including prolonged unemployment, divorce, and widowhood also increase the risk of substance abuse. Mental disorders are also risk factors of alcohol and drug abuse.

REFERENCES

[1] Miech R, Johnston L, O'Malley PM, Keyes KM, Heard K. Prescription opioids in adolescence and future opioid misuse. Pediatrics 2015;136:e1169—77.

[2] Ling W, Mooney L, Hillhouse M. Prescription opioid abuse, pain and addiction: clinical issues and implications. Drug Alcohol Rev 2011;30:300—5.

[3] Russell BS, Trudeau JJ, Leland AJ. Social influence on adolescent polysubstance use: the escalation to opioid use. Subst Use Misuse 2015;50:1325—31.

[4] Eitan S, Emery MA, Bates MLS, Horrax C. Opioid addiction: who are your real friends? Neurosci Biobehav Rev 2017;83:697—712.

[5] Munno D, Saroldi M, Bechon E, Sterpone SC, et al. Addictive behavior and personality traits in adolescents. CNS Spectr 2016;21:207—31.

[6] Hasin DS, O'Brien CP, Auriacombe M, Borges G, et al. DSM-5 criteria for substance use disorders: recommendations and rationale. Am J Psychiatr 2013;170:834—51.

[7] Grant BF, Saha TD, Ruan WJ, Goldstein RB, et al. Epidemiology of DSM-5 drug use disorder: results from the national epidemiologic survey on alcohol and related conditions-III. JAMA Psychiatr 2016;73:39—47.

[8] Brady KT, Randall CL. Gender differences in substance use disorders. Psychiatr Clin 1999;22:241—52.

[9] Kocherlakota P. Neonatal abstinence syndrome. Pediatrics 2014;134:e547—561.

[10] Sutter MB, Leeman L, Hsi A. Neonatal opioid withdrawal syndrome. Obstet Gynecol Clin N Am 2014;41:317—34.

[11] Klaman SL, Isaacs K, Leopold A, Perpich J, et al. Treating women who are pregnant and parenting for opioid use disorder and the concurrent care of their infants and children: literature review to support national guidance. J Addiction Med 2017;11:178—90.

[12] Patrick SW, Davis MM, Lehman CU, Cooper WO. Increasing incidence and geographic distribution of neonatal abstinence syndrome: United States 2009 to 2012. J Perinatol 2015;35:650—5.

[13] Whiteman VE, Salemi JL, Mogos MF, Cain MA, et al. Maternal opioid drug use during pregnancy and its impact on perinatal morbidity, mortality, and the costs of medical care in the United States. J Pregnancy 2014; 2014:906723.

[14] Soyka M. Buprenorphine use in pregnant opioid users: a critical review. CNS Drugs 2013;27:653—62.

[15] Tran TH, Griffin BL, Stone RH, Vest KM, Todd TJ. Methadone, buprenorphine, and naltrexone for the treatment of opioid use disorder in pregnant women. Pharmacotherapy 2017;37:824—39.

[16] Matthews TJ, MacDorman MF, Thoma ME. Infant mortality statistics from the 2013 period link birth/infant death data set. Natl Vital Stat Rep 2015;64:1—30.

[17] Broussard CS, Rasmussen SA, Reefhuis J, Friedman JM, et al. Maternal treatment with opioid analgesics and risk for birth defects. Am J Obstet Gynecol 2011;204. 314.e1—11.

[18] Yazdy MM, Mitchell AA, Tinker SC, Parker SE, Werler MM. Periconceptional use of opioids and the risk of neural tube defects. Obstet Gynecol 2013;122: 838—44.

[19] Lind JN, Interrante JD, Ailes EC, Gilboa SM, et al. Maternal use of opioids during pregnancy and congenital malformations: a systematic review. Pediatrics 2017;139:e20164131.

[20] Garrison L, Leeman L, Savich RD, Gutierrez H, et al. Fetal growth outcomes in a cohort of polydrug- and opioid-dependent patients. J Reprod Med 2016;61:311—9.

[21] Osborn DA, Jeffery HE, Cole M. Opiate treatment for opiate withdrawal in newborn infants. Cochrane Database Syst Rev 2005. https://doi.org/10.1002/14651858.

[22] Källén B, Reis M. Use of tramadol in early pregnancy and congenital malformation risk. Reprod Toxicol 2015;58: 246—551.

[23] Meyer M. The perils of opioid prescribing during pregnancy. Obstet Gynecol Clin N Am 2014;41: 297—306.

[24] Wilson GS, McCreary R, Kean J, Baxter JC. The development of preschool children of heroin-addicted mothers: a controlled study. Pediatrics 1979;63(1):135—41.

[25] Nygaard E, Moe V, Slinning K, Walhovd KB. Longitudinal cognitive development of children born to mothers with opioid and polysubstance use. Pediatr Res 2015;78: 330—5.

[26] Zedler BK, Mann AL, Kim MM, Amick HR, et al. Buprenorphine compared with methadone to treat pregnant women with opioid use disorder: a systematic review and meta-analysis of safety in the mother, fetus and child. Addiction 2016;111:2115—28.

[27] Wall AE, Kohl PL. Substance use in maltreated youth: findings from the national survey of child and adolescent well-being. Child Maltreat 2007;12:20—30.

[28] Singh VS, Thornton T, Tonmyr L. Determinants of substance abuse in a population of children and adolescents involved with the child welfare system. Int J Ment Health Addiction 2011;9(4):382—97.

[29] Kilpatrick DG, Acierno R, Saunders B, Resnick HS, et al. Risk factors for adolescent substance abuse and dependence: data from a national sample. J Consult Clin Psychol 2000;68:19—30.

[30] Tonmyr L, Thornton T, Draca J, Wekerle C. A review of childhood maltreatment and adolescent substance use relationship. Curr Psychiatr Rev 2010;6:223—34.

[31] Whitesell M, Bachand A, Peel J, Brown M. Familial, social, and individual factors contributing to risk for adolescent substance use. J Addiction 2013:579310.

[32] Alvarez-Alonso MJ, Jurado-Barba R, Martinez-Martin N, Espin-Martin N, et al. Association between maltreatment and polydrug use among adolescents. Child Abuse Negl 2016;51:379—89.

[33] Skinner ML, Hong S, Herrenkohl TI, Brown EC. Longitudinal effects of early childhood maltreatment on the co-occurring substance misuse and mental health problems in adulthood: the role of adolescent alcohol use and depression. J Stud Alcohol Drugs 2016;77:464—72.

[34] Canetti L, Bachar E, Galili-Weisstub E, De-Nour AK, Shalev AY. Parental bonding and mental health in adolescence. Adolescence 1997;32:381—94.

[35] Lei Y, Xi C, Li P, Luo M, et al. Association between childhood maltreatment and non-medical prescription opioid use among Chinese senior high school students: the moderating role of gender. J Affect Disord 2018;235: 421—7.

[36] Hussey JM, Chang JJ, Kotch JB. Child maltreatment and adolescent mental health problems in a large birth cohort. Child Abuse Negl 2013;37:292—302.

[37] Quinn K, Boone L, Scheidell JD, Mateu-Gelabert P, et al. Drug Alcohol Depend 2016;169:190—8.

[38] Barahmand U, Khazaee A, Hashjin GS. Emotion dysregulation mediates between childhood emotional abuse and motives for substance use. Arch Psychiatr Nurs 2016;30:653—9.

[39] Wolff S, Holl J, Stopsack M, Arens EA, et al. Does emotion dysregulation mediate the relationship between early maltreatment and later substance dependence? Findings of the CANSAS study. Eur Addiction Res 2016;22:292—300.

[40] Oshri A, Rogosch FA, Burnette M, Cicchetti D. Developmental pathways to adolescent cannabis abuse and dependence: child maltreatment, emerging personality, and internalizing versus externalizing psychopathology. Psychol Addict Behav 2011;25:634—44.

[41] Dube SR, Anda RF, Felitti VJ, Croft JB, et al. Growing up with parental alcohol abuse: exposure to childhood abuse, neglect, and household dysfunction. Child Abuse Negl 2001;25:1627—40.

[42] Appleyard K, Berlin LJ, Rosanbalm KD, Dodge KA. Preventing early child maltreatment: implications from a longitudinal study of maternal abuse history, substance use problems, and offspring victimization. Prev Sci 2011;12:139—49.

[43] Cicchetti D, Handley ED. Child maltreatment and the development of substance use and disorder. Neurobiol Stress 2019;10:100144.

[44] Elliott JC, Stohl M, Wall MM, Keyes KM, et al. The risk for persistent adult alcohol and nicotine dependence: the role of childhood maltreatment. Addiction 2014; 109:842—50.

[45] Taplin C, Saddichha S, Li K, Krausz MR. Family history of alcohol and drug abuse, childhood trauma, and age

of first drug injection. Subst Use Misuse 2014;49(10): 1311–6.

[46] Gerra G, Leonardi C, Cortese E, Zaimovic A, et al. Childhood neglect and parental care perception in cocaine addicts: relation with psychiatric symptoms and biological correlates. Neurosci Biobehav Rev 2009;33:601–10.

[47] Shin SH, Hong HG, Hazen AL. Childhood sexual abuse and adolescent substance use: a latent class analysis. Drug Alcohol Depend 2010;109:226–35.

[48] Ağaçhanlı R, Alnıak İ, Evren C. Sexual dysfunctions are predicted by childhood sexual abuse in women with opioid use disorder. Subst Use Misuse 2018;53:2184–9.

[49] Heffernan K, Cloitre M, Tardiff K, Marzuk PM, et al. Childhood trauma as a correlate of lifetime opiate use in psychiatric patients. Addict Behav 2000;25:797–803.

[50] Sansone RA, Whitecar P, Wiederman MW. The prevalence of childhood trauma among those seeking buprenorphine treatment. J Addict Dis 2009;28:64–7.

[51] Johnson JL, Leff M. Children of substance abusers: overview of research findings. Pediatrics 1999;103:1085–99.

[52] Jurich AP, Polson CJ, Jurich JA, Bates RA. Family factors in the lives of drug users and abusers. Adolescence 1985; 20:143–59.

[53] Tomčíková Z, Veselská ZD, Gecková AM, van Dijk JP, Reijneveld SA. Adolescents' drinking and drunkenness more likely in one-parent families and due to poor communication with mother. Cent Eur J Publ Health 2015;23:54–8.

[54] Richter M, Leppin A, Nic Gabhainn S. The relationship between parental socio-economic status and episodes of drunkenness among adolescents: findings from a cross-national survey. BMC Publ Health 2006;6:289.

[55] Melotti R, Heron J, Hickman M, Macleod J, et al. Adolescent alcohol and tobacco use and early socioeconomic position: the ALSPAC birth cohort. Pediatrics 2011; 127:e948–955.

[56] Van der Vorst H, Engles RC, Meeus W, Dekovic M. The impact of alcohol-specific rules, parental norms about early drinking and parental alcohol use on adolescent's drinking behavior. J Child Psychol Psychiatr 2006;47: 1299–306.

[57] Schiff M, Plotnikova M, Dingle K, Williams GM. Does adolescent's exposure to parental intimate partner conflict and violence predict psychological distress and substance use in young adulthood? A longitudinal study. Child Abuse Negl 2014;38:1945–54.

[58] Smith CA, Elwyn LK, Ireland TO, Thornberry TP. Impact of adolescent exposure to intimate partner violence on substance use in early adulthood. J Stud Alcohol Drugs 2010;71:2190230.

[59] Smith PH, Homish GG, Leonard KE, Cornelius JR. Intimate partner violence and specific substance use disorders: findings from the National Epidemiologic Survey on Alcohol and Related Conditions. Psychol Addict Behav 2012;26:236–45.

[60] Simons-Mortyon B, Haynie D, Liu D, Chaurasia A, et al. The effect of residence, school status, work status, and social influence on the prevalence of alcohol use among emerging adults. J Stud Alcohol Drugs 2016;77:121–32.

[61] Guo G, Li Y, Wang H, Cai T. Peer influence, genetic propensity, and binge drinking: a natural experiment and a replication. Am J Sociol 2015;121:914–54.

[62] Patrick ME, Schulenberg JE. Prevalence and predictions of adolescent alcohol use and binge drinking in the United States. Alcohol Res 2013;35:193–200.

[63] Stone AL, Carlisle SK. Racial bulling and adolescent substance use: an examination of school-attending young adolescents in the United States. J Ethn Subst Abuse 2015 [e-pub ahead of print].

[64] Ross CS, Maple E, Siegel M, DeJong W, et al. The relationship between brand specific alcohol advertisement on television and brand specific consumption among underage youth. Alcohol Clin Exp Res 2014;38: 2234–42.

[65] Waylen A, Leary S, Ness A, Sargent J. Alcohol use in films and adolescent alcohol use. Pediatrics 2015;135:851–8.

[66] Rostosky SS, Danner F, Riggle ED. Is religiosity a protective factor against substance abuse in young adulthood? Only if you're straight. J Adolesc Health 2007;40:440–7.

[67] Schulenberg J, Patrick ME, Maslowsky J, Maggs JL. The epidemiology and etiology of adolescent substance use in developmental perspective. In: Lewis M, Rudolph KD, editors. Handbook of developmental psychopathology. New York: Springer; 2014. p. 601–20.

[68] Whitesell M, Bachand A, Peel J, Brown M. Familial, social, and individual factors contributing to risk for adolescent substance use. J Addict 2013;2013:579310.

[69] Dakwar E, Levin FR. The emerging role of meditation in addressing psychiatric illness, with a focus on substance use disorders. Harv Rev Psychiatr 2009;17:254–67.

[70] Kissen M, Kissen-Kohn DA. Reducing addictions via the self-soothing effects of yoga. Bull Menninger Clin 2009; 73:34–43.

[71] Hallgren M, Romberg K, Bakshi AS, Andréasson S. Yoga as an adjunct treatment for alcohol dependence: a pilot study. Compl Ther Med 2014;22:441–5.

[72] Dhawan A, Chopra A, Jain R, Yadav D, et al. Effectiveness of yogic breathing intervention on quality of life of opioid dependent users. Int J Yoga 2015;8:144–7.

[73] Butzer B, LoRusso A, Shin SH, Khalsa SB. Evaluation of yoga for preventing adolescent substance use risk factors in a middle school setting: a preliminary group-randomized controlled trial. J Youth Adolesc 2017;46: 603–32.

[74] Raina N, Chakraborty PK, Basit MA, Samarth SN, Singh H. Evaluation of yoga therapy in alcohol dependence syndrome. Indian J Psychiatr 2001;43:171–4.

[75] Reddy S, Dick AM, Gerber MR, Mitchell K. The effect of a yoga intervention on alcohol and drug abuse risk in veteran and civilian women with posttraumatic stress disorder. J Alternative Compl Med 2014;20:750–6.

[76] Tang YY, Tang R, Posner MI. Mindfulness meditation improves emotion regulation and reduces drug abuse. Drug Alcohol Depend 2016;163(Suppl. 1):S13–8.

[77] Stephens MA, Wand G. Stress and the HPA axis: role of glucocorticoids in alcohol dependance. Alcohol Res 2012;34:468−83.

[78] Goeders NE. The impact of stress on addiction. Eur Neuropsychopharmacol 2003;13:435−6.

[79] Carroll ME, Lynch WJ, Roth ME, Morgan AD, Cosgrove KP. Sex and estrogen influence drug abuse. Trends Pharmacol Sci 2004;25:273−9.

[80] Rougemont-Bücking A, Grazioli VS, Marmet S, Daeppen JB, et al. Non-medical use of prescription drugs by young men: impact of potentially traumatic events and of social-environmental stressors. Eur J Psychotraumatol 2018;9:1468706.

[81] Smith KZ, Smith PH, Cercone SA, McKee SA, Homish GG. Past year non-medical opioid use and abuse and PTSD diagnosis: interactions with sex and associations with symptom clusters. Addict Behav 2016; 58:167−74.

[82] Keyes KM, Hatzenbuehler ML, Grant BF, Hasin DS. Stress and alcohol: epidemiological evidence. Alcohol Res 2012;34:391−400.

[83] Grieger TA, Fullerton CS, Ursano RJ. Posttraumatic stress disorder, alcohol use, and perceived safety after the terrorist attack on the pentagon. Psychiatr Serv 2003; 54:1380−2.

[84] Palijan TZ, Kovacevic D, Radeljak S, Kovac M, et al. Forensic aspects of alcohol abuse and homicide. Coll Anthropol 2009;33:893−7.

[85] Henkel D. Unemployment and substance use: a review of the literature (1990−2010). Curr Drug Abuse Rev 2011;4:4−27.

[86] Compton WM, Gfroerer J, Conway KP, Finger MS. Unemployment and substance outcomes in the United States 2002−2010. Drug Alcohol Depend 2014;142: 350−3.

[87] Bruguera P, Reynolds J, Gilvarry E, Braddick F, et al. How does economic recession affect substance use? A reality check with clients of drug treatment centres. J Ment Health Pol Econ 2018;21(1):11−6.

[88] Herttua K, Martikainen P, Vahtera J, Kivimäki M. Living alone and alcohol-related mortality: a population-based cohort study from Finland. PLoS Med 2011;8(9): e1001094.

[89] Labell LS. Wife abuse: a sociological study of battered women and their mates. Victimology 1979;4:258−67.

[90] Martin SL, Beaumont JL, Kupper LL. Substance use before and during pregnancy: links to intimate partner violence. Am J Drug Alcohol Abuse 2003;29:599−617.

[91] Coomber K, Mayshak R, Liknaitzky P, Curtis A, et al. The role of illicit drug use in family and domestic violence in Australia. J Interpers Violence 2019 [e-pub ahead of print].

[92] Kendler KS, Lönn SL, Salvatore J, Sundquist J, Sundquist K. Divorce and the onset of alcohol use disorder: a Swedish population-based longitudinal cohort and co-relative study. Am J Psychiatr 2017;174:451−8.

[93] Edwards AC, Larsson Lönn S, Sundquist J, Kendler KS, Sundquist K. Associations between divorce and onset of drug abuse in a Swedish national sample. Am J Epidemiol 2018;187:1010−8.

[94] Reboussin BA, Green KM, Milam AJ, Furr-Holden CD, Ialongo NS. Neighborhood environment and urban African American marijuana use during high school. J Urban Health 2014;91:1189−201.

[95] Cambron C, Kosterman R, Rhew IC, Catalano RF, et al. An examination of alcohol use disorder symptoms and neighborhood disorganization from age 21 to 39. Am J Community Psychol 2017;60:267−78.

[96] Kendler KS, Lonn SL, Salvatore J, Sundquist J, et al. Effect of marriage on risk of onset of alcohol use disorder: a longitudinal and co-relative analysis in a Swedish national sample. Am J Psychiatr 2016;173:911−8.

[97] Tenorio KA, Lo CC. Social Location, social integration, and the co-occurrence of substance abuse and psychological distress. Am J Drug Alcohol Abuse 2011;37: 218−23.

[98] Hearld KR, Badham A, Budhwani H. Statistical effects of religious participation and marriage on substance use and abuse in racial and ethnic minorities. J Relig Health 2017;56:1155−69.

[99] Debnam KJ, Milam AJ, Mullen MM, Lacey K, Bradshaw CP. The moderating role of spirituality in the association between stress and substance use among adolescents: differences by gender. J Youth Adolesc 2018;47:818−28.

[100] Treloar HR, Dubreuil ME, Robert Miranda JR. Spirituality and treatment of addictive disorders. R I Med J 2014; 97:36−8.

[101] McCree RR, John OP. An introduction to five factor model and its applications. J Pers 1992;60:175−215.

[102] Gore WL, Widiger TA. The DSM-5 dimensional trait model and five factor model of general personality. J Abnorm Psychol 2013;122:816−21.

[103] Widiger TA, Presnall JR. Clinical application of the five-factor model. J Pers 2013;81L:515−27.

[104] Martin ED, Sher KJ. Family history of alcoholism, alcohol use disorders and five factor model of personality. J Stud Alcohol 1994;55:81−90.

[105] Terracciano A, Lockenhoff CE, Crum RM, Bienvenu OJ, et al. Five factor model personality profiles of drug abusers. BMC Psychiatr 2008;8:22.

[106] Sutin AR, Evans MK, Zonderman AB. Personality traits and illicit substances: the moderating role of poverty. Drug Alcohol Depend 2013;131:247−51.

[107] Bozkurt M, Evren C, Cay Y, Evren B, et al. Relationship of personality dimension with impulsivity in alcohol dependent inpatients men. Nord J Psychiatr 2014;68: 316−22.

[108] Wingo T, Nesil T, Choi JS, Li MD. Novelty seeking and drug addiction in humans and animals: from behavior to molecules. J Neuroimmune Pharmacol 2016;11: 456−70.

[109] Liang CS, Ho PS, Yen CH, Chen CY, et al. Prog Neuro-Psychopharmacol Biol Psychiatr 2017;74:36−42.

[110] Smith RV, Young AM, Mullins UL, Havens JR. Individual and network correlates of antisocial personality disorder

among rural nonmedical prescription opioid users. J Rural Health 2017;33:198–207.

[111] Kienast T, Stoffers J, Bermpohl F, Lieb K. Borderline personality disorder and comorbid addiction: epidemiology and treatment. Dtsch Arztebl Int 2014;111:280–6.

[112] Smith JP, Book SW. Anxiety and substance use disorders: a review. Psychiatr Times 2008;25:19–23.

[113] Messer T, Lammers G, Müller-Siecheneder F, Schmidt RF, Latifi S. Substance abuse in patients with bipolar disorder: a systematic review and meta-analysis. Psychiatr Res 2017;253:338–50.

[114] Davis L, Uezato A, Newell JM, Fraizer E. Major depression and comorbid substance use disorder. Curr Opin Psychiatr 2008;21:14–8.

[115] Cassidy F, Ahearn EP, Carroll BJ. Substance abuse in bipolar disorder. Bipolar Disord 2001;3:181–4.

[116] Toftdahl NG, Nordentoft M, Hjorthoj C. Prevalence of substance abuse disorders in psychiatric patients, a nationwide Danish population based study. Soc Psychiatr Psychiatr Epidemiol 2016;51:129–40.

[117] Mauri MC, Volonteri LS, De Gaspari IF, Colasanti A, et al. Substance abuse in first episode schizophrenia patients: a retrospective study. Clin Pract Epidemiol Health 2006;23:4.

[118] Nielsen SM, Toftdahl NG, Nordentoft M, Hjorthøj C. Association between alcohol, cannabis, and other illicit substance abuse and risk of developing schizophrenia: a nationwide population based register study. Psychol Med 2017;47:1668–77.

[119] Khokhar JY, Dwiel LL, Henricks AM, Doucette WT, Green AI. The link between schizophrenia and substance use disorder: a unifying hypothesis. Schizophr Res 2018;194:78–85.

CHAPTER 6

Poppy Seed Defense and Workplace Drug Testing: Does It Work in the Court?

INTRODUCTION

Misuse and abuse of opioids as well as abuse of heroin is a serious public health issues. During legal drug testing including workplace drug testing, the presence of opiates (morphine, codeine, and heroin metabolite 6-monoacetylmorphine [6-MAM]) is tested in urine (most commonly) or other biological matrix including hair. In 2015, SAMHSA (Substance Abuse and Mental Health Services Administration) has added testing for oxycodone/oxymorphone as well as hydrocodone/hydromorphone in workplace drug testing. If opiate or opioid is confirmed in urine and no valid explanation is given by the person, then drug test should be considered as positive and may have negative consequence on that person including denying a job or being fired from a job.

If the workplace drug testing is analytical positive i.e., presence of codeine, morphine, or 6-monoacetylmorphine is confirmed by an alternative method such as gas chromatography-mass spectrometry (GC/MS) or liquid chromatography combined with mass spectrometry (LC-MS) or tandem mass spectrometry (LC-MS/MS), then person subjected to drug testing must provide a valid reason of positive drug testing such as taking prescription codeine or morphine. However, there is no defense for the presence of 6-MAM in urine because it is the marker of heroin abuse. In absence of a prescription for codeine or morphine, a very popular defense is eating poppy seed—containing foods. However, if the urine drug testing is positive for other opiates or opioids (except for morphine or codeine), then eating poppy seed—containing food is not a valid defense. Opiates and opioids inconsistent with eating poppy seed—containing foods are listed in Table 6.1.

Dihydrocodone is structurally related to codeine, but dihydrocodone is metabolized to dihydromorphone [1]. Therefore, prescription of dihydrocodone is not consistent with confirmation of morphine or codeine in urine. Therefore, poppy seed defense is inconsistent with dihydrocodone abuse.

DIFFERENTIATING HEROIN ABUSE FROM POPPY SEED DEFENSE

Because poppy seed defense can be used by illicit drug users, it is important to differentiate heroin abusers from people consuming poppy seed—containing food. Heroin has a very short plasma half-life (approximately 5 min) and is quickly metabolized to 6-MAM (also known as 6-acetylmorphine [6-AM]) by esterases in the liver, plasma, and erythrocytes. As a result, heroin may not be present in urine. Jenkins et al. showed that heroin can be detected in saliva up to 60 min after intravenous administration and even up to 24 h after smoking heroin base [2]. Identification of 6-AM, a specific metabolite of heroin, is considered to be definitive evidence of heroin use. Although 6-AM has been identified in oral fluid following controlled heroin administration, there is little information of identification of 6-AM in workplace drug testing. Presley et al. evaluated the prevalence of positive test results for 6-AM in 77,218 oral fluid specimens collected over a 10-month period (January—October 2001) from private workplace testing programs. Only morphine-positive oral fluid specimens were tested by gas chromatography combined with tandem mass spectrometry (GC-MS/MS) for 6-AM. A total of 48 confirmed positive morphine results were identified out of which 32 specimens (66.7%) were positive for 6-AM. Concentrations of 6-AM in oral fluid ranged from 3 to 4095 ng/mL. It has been speculated based on controlled dose studies of heroin administration, that ratio >1 of 6-AM/morphine in oral fluid is consistent with heroin use within the last hour before specimen collection. The confirmation of 6-AM in 66.7% of morphine-positive oral fluid specimens indicates that oral fluid testing for opioids may offer advantages over urine in workplace drug testing programs and in testing drugged drivers for recent heroin use [3].

Not detecting 6-AM in saliva or urine does not exclude heroin abuse because 6-AM also known as 6-MAM is eventually metabolized to morphine and is conjugated with glucuronic acid (morphine-3-glucuronide,

Fighting the Opioid Epidemic. https://doi.org/10.1016/B978-0-12-820075-9.00006-5

TABLE 6.1
Opiates or Opioids Confirmed in Urine, Which are Inconsistent With Eating Poppy Seeds—Containing Foods.

Opiates Inconsistent with Poppy Seed Defense	Opioids Inconsistent with Poppy Seed Defense
Oxycodone	Fentanyl
Oxymorphone	Alfentanil
Hydrocodone	Sufentanil
Hydromorphone	Remifentanil
Buprenorphine	Meperidine
Dihydrocodeine	Methadone
	Levorphanol
	Pentazocine
	Tramadol
	Tapentadol

major inactive metabolite, morphine-6-glucuronide, minor active metabolite). However, codeine present in poppy seed may also metabolized into morphine and abusing morphine will also cause confirmed detection of morphine in urine. Therefore, if both 6-MAM and morphine are confirmed in urine then heroin abuse can also be confirmed. Borriello et al. commented that 6-MAM is a biomarker of heroin abuse in both urine and blood with 100% sensitivity for urine and 95% sensitivity for blood [4]. If morphine is absent in urine but 6-MAM is confirmed in urine that is also consistent with heroin abuse. In one study the authors analyzed sporadic samples positive in the immunologic opiate screening test and later confirmed for 6-MAM but no morphine-3 glucuronide was detected. Out of 1923 urine specimens analyzed, 423 were positive for 6-MAM. In 32 (7.6%) of the samples 6-MAM was detected while the morphine-3-glucuronide concentrations were below cutoff (300 ng/mL) and in some cases even below the limit of detection (15 ng/mL). The 32 samples with this excretion pattern came from 13 different individuals, all but one with previously known heroin abuse. Eleven urine samples, nine containing morphine-3-glucuronide and 6-MAM and two with only 6-MAM, were also analyzed for the presence of heroin. In six samples, including the two with only 6-MAM, heroin was detected. There are several plausible explanations for these findings. The intake may have taken place shortly before urine sampling. High concentrations of heroin and 6-MAM may inhibit UGT 2B7 (Uridine 5'-diphospho-glucuronosyltransferase-2B7), the enzyme responsible for glucuronidation of morphine. Therefore, detection of 6-MAM in the absence of morphine-3-glucuronide also confirms heroin abuse [5].

Morphine Levels Indicative of Heroin Abuse

When no 6-MAM is present in urine, morphine concentration as well as morphine/codeine ratio may be used to suspect heroin abuse. Moriya et al. investigated urine drug levels of 440 specimens out of which 60 specimens were positive for opiates using EMIT (enzyme multiplied immunoassay technique). Out of 60 specimens 10 were judged to be collected from heroin abusers based on positive results for 6-MAM. In these specimens morphine-to-codeine ratios varied from 3.65 to 228. In one patient who was negative for codeine, the concentrations of free and total morphine were 0.114 and 2.22 μg/mL, respectively. In specimens 11 to 36 no 6-MAM was detected but both morphine and codeine were detected where morphine to codeine ratio was always greater than 1. Based on their data and literature data available, Moriya et al. proposed following criteria for heroin abuse when no 6-MAM is present in urine [6]:

- Detectable level of free morphine (unconjugated morphine in urine and concentration of total morphine is greater than 10,000 ng/mL) is present in urine.
- Detectable amount of codeine is present in urine.
- The morphine to codeine ratio is greater than two for both free and total amounts of morphine and codeine.

The authors further commented that after consumption of poppy seed—containing food, urine morphine concentrations do not seem to be higher than 300 ng/mL. Therefore, heroin abuse should be suspected if morphine concentration exceeds 1000 ng/mL in urine. Codeine users also show the presence of morphine in urine but morphine to codeine ratio remains much lower than one. However, in one urine specimen collected from a heroin abuser, the authors detected substantial amount of morphine but morphine/codeine ration was high indicating heroin abuse [6].

A morphine (M) to codeine (C) ratio greater than unity (M/C > 1) has been suggested as an indicator of heroin abuse. Konstantinova et al. examined the morphine to codeine ratio in a large population (N = 2438) of forensically examined autopsy cases positive for 6-MAM and/or morphine in blood and/or urine. Blood and urine concentrations of 6-MAM, morphine and codeine were examined using GC/MS as well as LC-MS/MS methods. The authors conclude that M/C ratio appeared to be a good marker of heroin use in postmortem cases. Both blood and urine M/C > 1 can be used to separate heroin users from other cases positive for morphine and codeine [7].

Ceder and Jones analyzed blood and urine specimens from 339 individuals suspected of driving under the influence of drugs as well as additional 882 blood specimens where urine specimens were not available. The authors identified 6-MAM in only 16 out of 675 blood specimens analyzed (2.3% positive for 6-MAM). In contrast, the authors detected 6-MAM in 212 out of 339 urine specimens analyzed (62% positive for 6-MAM). The authors concluded that only 2.3% heroin abusers were identified based on 6-MAM in blood but 62% heroin abusers could be identified using urine drug testing. In addition, the authors identified 85% of heroin abuser by using a morphine/codeine ratio above one in blood. The authors further commented that morphine/codeine ratio of <1 may indicate prescription use of codeine but not heroin abuse [8].

Heroin has a half-life of 2–6 min and is metabolized too quickly to be detected in autopsy samples. The presence of 6-AM (also known as 6-MAM) in urine, blood, or other samples is convincing evidence of heroin use by a decedent, but 6-AM itself has a half-life of 6–25 min before it is hydrolyzed to morphine. As a result, 6-AM may not be confirmed in postmortem specimen of a suspected heroin-related death. Codeine is often present in heroin preparations as an impurity and is not a metabolite of heroin. Studies report that a ratio of morphine to codeine greater than one indicates heroin use. In one study, the authors reviewed all accidental deaths investigation by the Jefferson County Coroner/Medical Examiner Office from 2010 to 13 where morphine was detected in postmortem blood. Five trauma related deaths were excluded from the review. In addition, any individual in whom no morphine or codeine was detected in a postmortem sample was excluded from further study.

Of the 230 deaths included in the analysis, 103 IV drug users with quantifiable morphine and codeine in a postmortem sample were identified allowing for calculation of an M/C ratio. In these IV drug users, the M/C ratio was greater than 1 in 98% of decedents. When controlling for the absence or presence of 6-AM, there was no statistically significant difference in the proportion of IV drug users when compared to non-IV drug users with an M/C ratio of greater than 1. The authors concluded that M/C ratio greater than 1 in an IV drug user is evidence of a death due to heroin toxicity even if 6-AM is not detected in the blood [9].

Instead of morphine to codeine ratio, some authors used codeine to morphine ratio in order to assess heroin abuse. If the total codeine to total morphine ratio in urine is < 0.5 and total morphine concentration in urine is > 200 ng/mL, codeine may be excluded as the source of morphine in urine. The likely source of morphine in such urine may be due to heroin abuse. Conversely, a total morphine concentration in urine <200 ng/mL

and a total codeine to total morphine ratio >0.5 indicates codeine ingestion [10].

Gambaro et al. analyzed postmortem samples from 14 cases of suspected heroin overdose and reported that the concentration of morphine in blood ranged from 33 to 688 ng/mL, while the concentration of codeine ranged from 0 to 193 ng/mL. However, in the brain, the concentration of morphine was found to be between 85 and 396 ng/g, while the levels of codeine ranged from 11 to 160 ng/g. The codeine/morphine ratio in the blood ranged from 0.043 to 0.619; however, in the brain, the same ratio was found to be between 0.129 and 0.552. Therefore, codeine to morphine ratio may exceed 0.5 in blood of some heroin abusers. However, codeine to morphine ratio in urine was always significantly lower than 0.5 in all heroin abusers [11].

Codeine is not classified as prohibited drug by the world antidoping agency, but morphine, a metabolite of codeine, is a prohibited drug during competition. Metabolism of codeine to morphine is also affected by polymorphism of *CYP2D6* gene encoding CYP2D6 enzyme. Urine sample of a soccer player tested positive for morphine, but the player denied taking any narcotics including morphine but admitted taking several acetaminophen codeine tablet (300 mg acetaminophen, 10 mg codeine) day before drug testing due to tooth pain. The observed morphine to codeine ratio was 1.03 in urine. The disciplinary committee concluded morphine in his urine was due to codeine intake and was cleared of any wrong doing. The disciplinary committee's decision was based on the fact that the athlete could be an ultrarapid CYP2D6 metabolizer where morphine/codeine ratio may slightly be above 1 (in this case 1.03) [12]. Although after codeine intake morphine/codeine ratio in blood or urine should be below 1, in one study, the authors showed that in ultrarapid metabolizer, morphine to codeine ratio in blood may slightly exceeds the value of 1 (highest value 1.060) when codeine is consumed approximately 24 h ago but the ratio should be below 1 when codeine is taken 12 h before the testing [13].

Acetylcodeine

A major limitation of 6-MAM is its short half-life and may often be absent in urine collected from heroin abusers due its short window of detection. Therefore, acetylcodeine has been suggested as a marker for heroin abuse. Codeine is present naturally in opium, which is converted into acetylcodeine, a synthetic by-product during acetylation of morphine to produce illicit heroin. However, acetylcodeine is not a better marker than 6-MAM for heroin abuse due to shorter window of detection in urine compared to 6-MAM as well as substantially lower concentrations than 6-MAM. In one study, the authors analyzed 100 criminal justice urine specimens containing >5000 ng/mL of morphine and detected acetylcodeine only in 37 specimens with concentrations ranging from 2 ng/mL to 290 ng/mL (median: 11 ng/mL). However, 6-MAM was also present in these specimens with much higher concentrations than acetylcodeine (range: 49−12, 600 ng/mL; median: 740 ng/mL). Of 63 specimens negative for acetylcodeine, 36 specimens showed the presence of 6-MAM at concentrations ranging from 12 ng/mL to 4600 ng/mL (median: 124 ng/mL). The authors commented that when detected, acetylcodeine concentration was much lower than 6-MAM (average 2.2% of 6-MAM concentration). Therefore, due to very low concentration of acetylcodeine in urine of heroin abusers, this marker is less reliable than 6-MAM to confirm heroin abuse in workplace drug testing or criminal justice system. However, if present in urine, detection of acetylcodeine further validates heroin abuse [14]. In another study, the authors reported that average half-life of acetylcodeine in urine is 237 min and as a result peak acetylcodeine concentration is observed 2 h after administration and the detection window is only 8 h [15].

Acetylcodeine is also present in oral fluid. Phillips and Allen analyzed 513 oral fluid specimens using liquid chromatography linked to atmospheric pressure ionization tandem mass spectrometry (LC-MS/MS) and observed detectable amounts of one or more opiates in 297 specimens. Out of these specimens morphine, codeine, 6-MAM, acetylcodeine, and heroin were detected in 97%, 82%, 77%, 55%, and 45%, respectively. However, nine specimens showed detectable amount of heroin but no acetylcodeine. The authors concluded that detection of acetylcodeine in oral fluid indicates heroin abuse but has limited application [16].

Papaverine and Metabolites

Papaverine (1-(3,4-dimethoxybenzy1)-6,7-dimethoxyisoquinoline), an opium alkaloid, is sometimes found in illicit heroin preparations. Papaverine has a half-life of 0.8−1.5 h and is metabolized to a large extent by demethylation, the mono- and didemethylated metabolites being excreted as glucuronides and sulfates conjugates. Approximately 50% of the metabolites of papaverine are excreted in the urine within 48 h, 6-desmethylpapaverine being the major metabolite in the urine. Therefore, 6-desmethylpapaverine and its glucuronide conjugate have been suggested to be potential biomarkers for illicit heroin abuse since they have longer detection windows than 6-MAM in urine

samples. However, major limitation of using papaverine as urinary biomarker of heroin abuse is that papaverine can also be found in medicines for the treatment of erectile dysfunction and in patients with histories of tracheal intubation for which neuromuscular relaxant atracurium was administered because atracurium undergoes Hofmann elimination with laudanosine as one of the major metabolites, which is then partially dehydrogenated to papaverine [17].

ATM4G: a Biomarker of Heroin Abuse

More recently, ATM4G (acetylated-thebaine-4-metabolite glucuronide), a new marker for the discrimination between street heroin consumption and poppy seed ingestion, has been described by Chen et al. This new biomarker of heroin abuse is due to the presence of thebaine in opium which is acetylated during illicit manufacture of illicit heroin. Then this acetylated product undergoes molecular rearrangement. After heroin abuse, this product is metabolized into ATM4G. When 22 specimens collected from heroin abusers which were tested negative for 6-MAM by GC/MS (reporting threshold: 10 ng/mL) were analyzed by LC-MS/MS, peak corresponding to ATM4G was identified in 16 out of 22 specimens. In contrast, 6-MAM was detected by LC-MS/MS only in three specimens with concentration >1 ng/mL. The authors concluded that ATM4G is a novel urine biomarker to identify heroin abusers [18].

Mass et al. commented that discrimination between street heroin consumption and poppy seed ingestion represents a major toxicological challenge in daily routine work. However, a novel opportunity to overcome these hindrances is represented by the new potential street heroin marker ATM4G, originating from thebaine during street heroin synthesis followed by metabolic reactions after administration. In the study by Mass et al., urine samples after consumption of different German poppy seed products and urine samples from subjects with suspicion of preceding heroin consumption were tested for ATM4G, 6-acetylcodeine, papaverine, noscapine, 6-MAM, morphine, and codeine. As expected, after consumption of poppy seeds, only morphine and codeine could be detected in urine specimens but no acetylcodeine, 6-MAM, or ATM4G were detected. In addition, neither papaverine nor noscapine could be detected in urine specimens even after consumption of poppy seeds containing up to 37 µg noscapine and up to 9.8 µg papaverine, respectively. In contrast, when urine specimens collected from suspected heroin abusers were analyzed, ATM4G could be detected in 9 out of 43 cases, but 6-MAM and acetylcodeine were present in only 7 urine specimens. The authors concluded that ATM4G could be detected in urine samples in which neither 6-MAM nor 6-acetylcodeine could be detected. The authors concluded that ATM4G is a specific biomarker of street heroin abuse [19]. Biomarkers of heroin abuse are listed in Table 6.2.

OPIATE CONTENT OF POPPY SEEDS

In general, morphine is the most abundant alkaloid found in raw opium (4%–21%), followed by noscapine (4%–8%), codeine (0.8%–2.5%), papaverine (0.5%–2.5%), and thebaine (0.5–2) [20]. Morphine, codeine, and thebaine are found in the opium poppy latex. Opium alkaloids are not found in poppy seeds but are transferred onto the seed coats during harvesting process. As a result, alkaloid content of poppy seeds varies widely with morphine content ranging from 0.1 to 620 mg/kg of seeds and codeine content varying from 0.08 to 57.1 mg/kg of seeds. In general, morphine content is significantly higher than codeine [20]. Pelders and Ros analyzed poppy seeds from seven different origins (Dutch, Australian, Hungarian, Spanish, Czech, and two Turkish) and observed wide variation of morphine (2–251 µg/g) and codeine (0.4–57.1 µg/g) content. The authors found highest amount of morphine and codeine in poppy seeds imported from Spain [21].

The wide variations among morphine and codeine content of poppy seeds are related to type of poppy plant, geographical location, and time of harvest as well as processing. However, morphine and codeine content of poppy seeds are significantly lower compared to their content in opium. Traditionally ripe seeds are manually shaken so that seeds fall out of the holes below the many layered stigma. However, when seeds are machine processed, seeds may get contaminated with unripe capsules which are removed later. Moreover, washing the poppy seeds can significantly reduce the morphine content indicating that morphine is located on the surface of the seed [22]. Trafkowski et al. reported that the morphine content varied from less than 0.1–450 mg/kg of seed while codeine content varied from less than 0.1–57.1 mg/kg of seed. The thebaine content varied from 0.3 to 41 mg/kg of seed, while noscapine and papaverine content varied from 0.84 to 230 mg/kg and 0–67 mg/kg, respectively [23].

Report from the American Spice Trade Association, New York, indicated that Australian, Dutch, and Turkish poppy seeds represent approximately 94% of the US market of poppy seeds. The morphine content of the

TABLE 6.2
Biomarkers of Heroin Abuse Which are Inconsistent With Poppy Seed Defense.

Biomarker of Heron Abuse	Recommended value/ Comments
6-Monoactetylmorphine (6-MAM), also known as 6-acetylmorphine	Any level at or above the cutoff concentration of 10 ng/mL in urine confirms heroin abuse. However, 6-MAM can be detected in urine only 8–12 h after abuse. Confirmation of 6-MAM is inconsistent with poppy seed defense. However, absence of 6-MAM does not rule out heroin abuse. High morphine level in urine is consistent with heroin abuse.
Acetylcodeine	Any concentration that can be confirmed is consistent with heroin abuse and inconsistent with poppy seed defense. However, acetylcodeine like 6-MAM has a short half-life and may not be detected in urine of all heroin abusers.
Morphine	>2000 ng/mL is indicative of heroin abuse and inconsistent with poppy seed defense.
Morphine/codeine ratio	>1 in either blood or urine rules out poppy seed defense.
Papaverine, 6-desmethylpapaverine, and its glucuronide	Papaverine is present in opium from which heroin is produced. As a result, papaverine and its metabolites may be used as heroin biomarkers, and these compounds have longer detection window that 6-monoacetylmorphine. Major limitation is that papaverine may be present in individuals taking various medication.
ATM4G (acetylated-thebaine-4-metabolite glucuronide)	The confirmed presence of ATMG4 confirms heroin abuse and inconsistent with poppy seed defense.

Australian poppy seeds ranges from 90 to 200 µg of morphine per gm of poppy seed, while Dutch and Turkish poppy seed contain only 4–5 µg of morphine per gram of poppy seed [24]. Thevis et al. analyzed morphine and codeine content of eight different brands of poppy seeds available in the US market and reported that typical morphine content of poppy seeds varied from 0.6 microgram/g of poppy seed to 151.6 microgram/g of poppy seed [25].

Although morphine content of the Netherland poppy seeds is low, another report indicated that morphine content was 100 mg/kg of poppy seeds [26]. Zentai et al. analyzed 737 poppy seed samples in Hungary and detected morphine in 726 out of 737 specimens. Codeine and thebaine were detected in 61.3% and 63.0% of samples analyzed, while noscapine was present in only 6.2% of specimens analyzed. Interestingly, average morphine and codeine content were 18.7 mg/kg and 3.6 mg/kg, but highest morphine content was 533.0 mg/kg while highest codeine content was 60.0 mg/kg, indicating wide variation of morphine and codeine content in various specimens of poppy seeds [27].

MORPHINE AND CODEINE IN POPPY SEED–CONTAINING FOODS

When poppy seeds are incorporated into food, a substantial amount of morphine is degraded during food processing. As a result, a person consumes much less amount of morphine and codeine when eating breads, muffins, pastries, and cakes containing poppy seeds because most morphine is degraded during high baking temperature. As a result, intoxication or impairment after eating poppy seed–containing food has not been reported. Poppy seeds may also be used in salad dressing. Because both morphine and codeine are present in poppy seeds, it is expected that both morphine and codeine should be present in urine of a subject after eating poppy seed–containing food. In response to the presence of morphine and codeine in urine after consuming poppy seed–containing foods, the immunoassay cutoff for the Federal Government mandated drug testing was increased from 300 ng/mL to 2000 ng/mL to avoid false positive test results. However, in SAMHSA guidelines, heroin metabolite 6-MAM has a cutoff of 10 ng/mL, so that a heroin addict should not be able to pass a drug test. However, some private employers still use 300 ng/mL cutoff for opiate immunoassays in workplace drug testing. Eating poppy seed–containing foods and poppy seed containing salad dressing may cause false positive opiate test result

at 300 ng/mL cutoff [28]. In addition, few products contain very high amount of morphine due to use of excessive amounts of poppy seeds. Drinking or eating such product may cause opiate toxicity or even fatality. Such products must be avoided at all the time.

Toxicity and Fatality after Drinking Poppy Tea

Although growing poppy plant is illegal in the United States, buying poppy seeds is legal. In general, morphine and codeine content is much higher in unwashed and unprocessed poppy seeds which can be purchased in bulk from online sources legally. As a result, unwashed poppy seeds are highly desirable to drug abusers because they can extract morphine and codeine from poppy seeds. Poppy tea can be easily made by boiling poppy seeds with hot water where morphine and codeine present in poppy seeds are extracted into boiling water. However, poppy tea can also be prepared by boiling poppy stems with hot water. Drinking poppy tea may cause opiate addiction [29].

Braye et al. investigated abuse of poppy seed tea in New Zealand using 24 opiate-dependent patients attending the clinic. A total of 11 out of 24 patients (46%) reported drinking poppy seed tea. Poppy seed tea preparation took 1–2 h and involved soaking poppy seed, use of hot water, use of acidity (citric acid or lemon juice), reducing volume by boiling, and adding food flavored powder. The amount of poppy seeds used varied between 0.02 and 3.0 kg, but most people used 0.5–1.0 kg of poppy seeds for preparing tea. The patients reported a median onset of effect of 15 min (range 5–60 min) after ingestion of poppy seed tea and the effects lasted between 12 to more than 24 h. Psychological effects included calmness, euphoria, and other symptoms of opiate abuse. The authors concluded that poppy seed tea is commonly used by opiate-dependent patients attending an alcohol and drug rehabilitation clinic. Poppy seed tea is used because of its low cost, legal availability, and oral administration [30].

Nanjayya et al. described a case of an 82-year-old woman who was dependent on poppy tea for approximately 55 years. As access to poppy tea became more problematic in India due to legal restriction, she was brought by her family member to an addiction recovery center for treatment for her severe withdrawal symptoms. She was successfully treated on buprenorphine maintenance therapy [31]. Seyani et al. described a case of a 64-year-old woman who was admitted to the hospital due to respiratory arrest after drinking poppy tea. A naloxone infusion was able to reverse her opiate toxicity and she made good recovery. It was subsequently discovered that she had brewed tea from poppy buds which she picked from a nearby commercial poppy farm. She learnt that practice of preparing tea from poppy buds when she was in Afghanistan [32].

Significant amounts of morphine may be found in blood and urine of subjects consuming poppy tea. A baker consumed poppy tea prepared from seeds and experienced tonic-clonic seizure and delirium. His business partner informed that he was purchasing 25 kg of poppy seeds per week, whereas only 3 kg were required for bakery. The concentration of morphine in his blood was almost 3.0 mg/L. The patient admitted drinking about 2 L of poppy tea made from 4 kg of seeds. It was estimated that he was consuming approximately 280 mg morphine a day [33]. Van Thuyne et al. reported that the morphine content of poppy tea prepared from two specimens of a different species of poppy (*Papaveris fructus*) contained 10.4 (tea A) and 31.5 µg/mL(B) morphine. After administration of tea A in five healthy volunteers, the maximum morphine level of morphine was 4.34 µg/mL (4300 ng/mL), but after consuming tea B, the maximum urinary concentration of morphine was 7.4 µg/mL (7400 ng/mL) [34].

Powers et al. reported two fatalities that may be associated with consuming poppy tea. A 21-year-old white college student was found dead on his bed by his roommate. He had foam at his nose and mouth but there was no injury. The decedent and his fried would go to the grocery store and buy a bulk bag of poppy seeds. They would put seeds in water followed by shaking and then drank the water. Postmortem analysis of the victim showed the presence of morphine (>0.80 mg/L) and codeine (0.26 mg/L). Vitreous contained 0.46 mg/L of morphine and 0.26 mg/L of codeine. In another case report, a 24-year-old white male died from morphine intoxication. The decedent's blood showed 0.25 mg/L of morphine and 0.012 mg/L of codeine during autopsy. A trace amount of thebaine was also found in the blood. A 5-lb bag of commercially available poppy seeds was found in the scene along with a 33-fL oz bottle filled with poppy seeds and water. It was suspected that the decedent consumed a lethal amount of morphine from home-brewed poppy seed tea. The authors also purchased 22 poppy sample from online sources: bulk poppy seeds (n = 19), poppy seed powder (n = 1), poppy seed tea bag (n = 1), and liquid poppy extract (n = 1). For bulk poppy seeds, 150 mL of water was added to 85 gm of poppy seeds. For poppy seed tea bag, 30 mL of water was added to contents of two tea bags (6 gm). For poppy seed powder, 100 mL water was added to 35 gm of powder. For the liquid poppy extract, 13 mL of water was added to 7 mL of extract. After stirring at room temperature over a 10 min period, poppy seed tea were prepared using water preheated at 94°C to mimic brew recipes found online by the authors. The authors also used water acidified with lemon juice for extraction under acidic condition to mimic taste modifiers described on forums and tea review. Based on results, the authors concluded that regardless of extraction protocol, lethal amounts of morphine can be rinsed from poppy seed coats by home-brewing methods [35]. Steentoff et al. also reported seven deaths in Denmark during the period of 1982−85 due to consumption of opium tea [36].

Amazon, eBay, and Etsy, major online stores serving customers within and outside of the United States, offer poppy seeds for sale. A popular large health and wellness site (Mercola.com) recommends poppy seed tea for pain control and recipes appear there and on chewtheworld.com (neither of these sites sells poppy seeds). The website foodtolive.com sells different varieties of poppy seeds (identified by origin: English, Australian, Spanish) for $9.99 to $17.98 a pound. The website poppyseedtearecipe.com advertises poppy seeds from California as a "homemade high" and offers a pound of unwashed poppy seeds for $25.99. The website thenoddingturtle.biz sells European and Tasmanian "unwashed" poppy seeds with prices available upon e-mail request, as discounts may apply if a buyer opts for 10 lb or more. Poppy seeds are also sold on this site as an exfoliant. Sincerelynuts.com sells a variety of foodstuffs, including poppy seeds for $11.99 lb (English), $7.99 a pound (Dutch), or $4.99 a pound (Spanish). This site does not discuss poppy seed tea. Opiumpoppyseeds.com sells opium seeds to grow poppies [37]. As a result, both poppy seeds and recipe for making poppy seed tea are easily available to a lay person who may start consuming poppy seed tea without being awarded of its potential health hazards.

Ethnic Poppy Seed Foods

Poppy seed paste is a specific traditional food in Turkey and is made from roasted and twice grinded poppy seeds. Morphine content of different colored poppy seeds is variable from all over the world (0.1−294 mg/kg). It was reported that morphine content of Turkish origin poppy seeds is ranged between 9.7 and 37.4 mg/kg. In one study, urine opiates immunoassay results above the 300 ng/mL cutoff value were found in 63.4% of subjects after consuming white poppy seed paste, 73.4% after consuming yellow poppy seed paste, and 68.8% after consuming blue-black poppy seed paste. However, using 2000 ng/mL cut-off, the authors observed 21.1% positive specimens after ingesting white poppy seed paste, and 29.9% after ingesting blue-black poppy seed paste. According to cutoff value 300 ng/mL, opiate concentrations were found positive up to 48 h. For cutoff value 2000 ng/mL, this time was up to 12 h in collected urine samples after consumption of three different colored poppy seed pastes. In all urine samples, thebaine was detected while the heroin abuse metabolite 6-AM was not. Urine drug testing legislation was revised on 2016 in Turkey and opiate screening cutoff values increased from 300 to 2000 ng/mL. Overall results have shown that poppy seed paste as food consumption could lead to opiate positive urine test result even if increased cutoff levels are used. However, in the absence of detecting 6-AM, eating poppy seed paste as the reason positive opiate drug test at 2000 ng/mL is a valid reason [38].

Consuming Raw Poppy Seeds

There is an isolated report in the literature where a 54-year-old woman with intractable vomiting was found unresponsive at home and later pronounced dead. At autopsy, a cast-like large bowel obstruction composed of poppy seeds was identified. However, postmortem blood morphine level was <10 ng/mL. Cause of death was determined to be complications of a bowel obstruction secondary to poppy seed ingestion rather than morphine overdose [39].

EXPECTED MORPHINE AND CODEINE LEVELS AFTER EATING POPPY SEED—CONTAINING FOODS

Many investigators studied morphine and codeine concentrations in urine after subjects consumed either poppy seeds or poppy seed—containing food. A wide variation of both morphine and codeine levels have been reported by various investigators and such differences are related to whether subjects consumed raw poppy seeds or poppy seed—containing food. In general, both morphine and codeine concentrations are significantly higher after eating raw poppy seed compared to eating poppy seed—containing food because a substantial amount of morphine is destroyed during baking process of poppy seed cakes or muffins due to very high temperature during baking.

Degradation of Morphine During Food Processing

Morphine content of poppy seed is significantly reduced during food processing. Even simple washing can significantly remove morphine from poppy seeds. Studies have shown that washing poppy seeds with hot water is more effective than washing with cold water because washing with hot water (60°C) may reduce morphine content by approximately 70%. However, boiling water is not more effective than hot water for reducing morphine content of poppy seeds. Longer washing time can further reduce morphine content of poppy seeds. In addition to washing, drying of poppy seeds as well as grinding also reduces morphine content. The optimal treatment of poppy seed consists of washing, drying, and grinding prior to baking into foods. This process significantly reduces the morphine content and simultaneously improves the organoleptic quality of the product. Baking poppy seed containing cakes and buns also further reduces morphine content [40].

When a person eats poppy bun, poppy muffin, or poppy cakes, the person consumes only about 3 g of poppy seeds which is not a health risk because morphine content is significantly reduced during food processing. For example, in one study, the authors reported that median morphine content of poppy seeds was 6.8 mg/kg, which was reduced to 3.9 mg/kg during grinding and heating steps. Finally, in the finished food sold in bakeries, the morphine content was less than 1 mg/kg. Food processing including baking is capable of reducing morphine content of poppy seeds by 90% [22].

Simple grinding can reduce of morphine content by 34%. In addition, Sproll et al. also observed that for poppy cakes, only 16%—50% of morphine and 10%—50% of codeine present in original poppy seeds can be recovered but for poppy buns baked at high temperature (220°C), only 3% of morphine and 7% of codeine present in original poppy seeds was found. The authors concluded that baking temperature has a significant influence on reducing morphine content of poppy seeds because observed reduction of morphine at baking temperature of 135°C was relatively low (around 30%), but when baking temperature was 220°C, higher reduction in morphine content may be observed. The authors advised consumers and bakers to wash poppy seeds with water before direct use or before grinding seeds prior to baking. Moreover, it is advisable to use high temperature for baking in order to reduce morphine and codeine content in finished products. It is possible to remove 100% of morphine by superior food processing technique including baking at high temperature [20]. Morphine and codeine are found in poppy seeds both as free and bound form. In one study, the authors observed that washing removed 45.6% morphine and 48.4% of free codeine [41]. Most effective and less effective ways of reducing morphine content of poppy seeds during food processing are summarized in Table 6.3.

TABLE 6.3
Most Effective and Less Effective Way of Reducing Morphine Content of Poppy Seeds During Food Processing.

Food Processing	Morphine Reduction
Washing poppy seeds at 60°C followed by baking at 220°C	Most effective way of removing morphine from poppy seed—containing food where over 90% of morphine could be lost (removing by washing and degradation during high-temperature baking)
Baking food at 220°C	Very effective way of morphine reduction where morphine content reduced by 80%—90%
Washing with hot water (60°C)	Effective way of removing morphine because average loss is 70%
Baking food at 150°C	Less effective because only average loss of morphine is 30%
Grinding poppy seeds in poppy mills	Less effective because only average loss of morphine is 34%

In general, morphine concentrations are higher than codeine concentration after eating poppy seed—containing foods. Wide range of morphine concentrations (up to 13,857 ng/mL) has been reported by authors in volunteers after consuming various amounts of poppy seeds or poppy seed products. However, most of these studies were not controlled for food processing—related loss of morphine and codeine in poppy seeds. In some studies, volunteers consumed equivalent amount of unprocessed poppy seeds expected to be present on poppy seed—containing foods. However, such assumption is wrong because in real-life situation, morphine content should be significantly reduced in poppy seed cakes, bagels, or muffin because most morphine should be degraded during food processing.

High Morphine Levels After Eating Unprocessed Poppy Seeds

Most of the earlier studies have reported very high morphine content in subjects after eating unprocessed poppy seeds. Hill et al. investigated the effect of ingesting a high amount of poppy seed on the urinary concentrations of morphine in volunteers as well as concentrations in hair. The poppy seed study was performed using Australian poppy seed because it contains the highest amount of morphine of any poppy seed available in the US market. Ten subjects (six male, four female) ingested 150 g of poppy seeds over a period of 3 weeks. Urine specimens were collected on days when volunteers consumed poppy seeds while hair specimen was collected in the 5th week of the study. The range among the 10 subjects of the highest urine value for each subjects was 2929 to 13,857 ng/mL for morphine and 208—1174 ng/mL for codeine (determined by GC/MS). Moreover, urinary morphine levels remained above the 2000 ng/mL for as long as 10 h. Hair morphine levels were 0.05—0.48 ng/10 mg of hair [24]. However, these high morphine levels are unrealistic because during food processing most morphine should be destroyed.

Morphine Levels After Eating Poppy—Containing Foods

More recent publications indicate much lower morphine and codeine level after eating poppy seed—containing food, thus confirming that eating poppy seed—containing food should not cause positive opiate screen at 2000 ng/mL cutoff level. However, some private employers still use old 300 ng/mL cutoff for opiate immunoassays. As a result, a person may be tested positive for opiate screen after eating poppy seed—containing food.

ElSohly designed an experiment to determine urinary morphine and codeine level after realistic consumption of poppy seed—containing food. Two male and two female volunteers participated in four protocols of eating poppy seed containing rolls and each protocol was separated at least by 1 week. Subjects ingested one, two, or three poppy seed rolls each containing 2 g of Australian poppy seeds (108 mg morphine/kg of seed). In the fourth protocol, subjects ingested two rolls per day for four consecutive days. The highest concentration of morphine was observed 3—8 h after ingestion or in the first void specimen. Out of 264 urine specimens collected from these volunteers, only 16 specimens showed positive by immunoassay at 300 ng/mL cutoff. However, out if these 16 specimens, only 3 specimens showed morphine levels exceeding 400 ng/mL (406 ng/mL, 611 ng/mL, and 954 ng/mL) by GC/MS. However, all these positive specimens (300 ng/mL cutoff using immunoassays) tested negative within 24 h with opiate levels less than 150 ng/mL. When one volunteer consumed a poppy seed cake containing very high amount of (15 gm) of poppy seed, highest morphine level of 2010 ng/mL (codeine level 78 ng/mL) was observed 9 h postingestion [42]. This study clearly shows that eating realistic amount of poppy seed containing bake food such as roll, muffin, cake, etc., should not cause positive opiate test result at 2000 ng/mL cutoff except for unusual cases of eating food containing very high amounts of poppy seeds.

In another study, four volunteers ate three poppy seed bagels each. Neither morphine nor codeine was detected in oral fluids (limit of detection 3 ng/mL). However, trace amounts of codeine were detected in urine specimens and highest morphine concentration of 603 ng/mL was observed 1–4 h postingestion. Lowest morphine level was 314 ng/mL. These data also indicate that after eating poppy seed bagels, urine opiate levels were significantly below 2000 ng/mL cutoff. However, in the same study, authors also showed that eating unprocessed poppy seeds along with poppy seed bagel could produce very high morphine level in urine. When three volunteers ate one poppy seed bagel (820 mg of poppy seed) and then ingested an unlimited amount of poppy seeds in 1 hour (volunteer 1 ingested 14.82 g seeds, volunteer 2 ingested 9.82 g seeds, and volunteer 3 ingested 20.82 g seed), the urine morphine level in volunteer 3, exceeded 10,000 ng/mL approximately 2.4 h after ingestion. Urine specimens remained positive up to 8 h at 2000 ng/mL. In addition, the oral fluid tested positive up to 1 hour after ingestion of poppy seed bagels and poppy seeds at a 40 ng/mL cutoff (highest morphine: 205 ng/mL). This study indicated that eating poppy seed bagels may cause positive opiate screening at 300 ng/mL cutoff but not at 2000 ng/mL cutoff. However, eating raw poppy seeds along with bagel may cause very high urine morphine levels [43].

The study by Samano et al. also reached the same conclusion that urine as well as oral fluid morphine and codeine levels should be significantly lower after eating poppy seed–containing food compared to ingesting raw poppy seeds. In first part of the experiment, 12 volunteers consumed poppy seed roll, and in the second experiment, the same individuals consumed equivalent amount of raw poppy seeds containing approximately 3.3 mg of morphine and 0.6 mg of codeine in each dose. As expected, urinary morphine levels varied from 155 to 1408 ng/mL (as determined by GC/MS) after eating poppy seed rolls but much higher urine morphine levels were observed (294–4213 ng/mL) when same subjects in a later experiment consumed equivalent amount of raw poppy seeds. In addition, urinary codeine concentrations were significantly lower (range: 140–194 ng/mL) after eating

poppy seed rolls compared to after eating raw poppy seeds (121–664 ng/mL). Similar trend was also observed with oral fluids with much lower morphine and codeine levels after consuming poppy seed rolls (morphine: 7–143 ng/mL up to 1.5 h, codeine detected only in 5.5% specimens) compared to eating raw poppy seeds (morphine: 7–600 ng/mL, codeine: 8–112 ng/mL; 0.25–3 h after consumption). Therefore, it is very unlikely that a person would test positive using opiate immunoassay at 2000 ng/mL cutoff after eating poppy seed–containing food. However, at a lower cutoff of 300 ng/mL, some individuals may test opiate positive depending at what time urine specimen was collected after eating poppy seed–containing food. In general, highest morphine levels are observed between 3 and 8 h postingestion [44].

Because amount of morphine is very low in cooked poppy seed contain foods impairment is unlikely after eating such foods. Meneely reported that seven volunteers who ingested 25 g of poppy seed each baked into cakes showed opiate positive urine at 300 ng/mL shortly after consuming poppy seed products, but none of the volunteer exhibited symptoms of opiate impairment based on series of standardized drug recognition evaluation test [45]. Lo and Chua reported that when volunteers ingested a curry meal or two containing various amounts of washed seeds (morphine intake: 200.4–1002 µg; codeine intake: 95.9–479.5 µg), the urinary morphine levels were found to be in the range 120–1270 ng/mL urine and codeine level varied 40–730 ng/mL [41]. Therefore, eating such meal would cause positive opiate test result at 300 ng/mL.

In Germany, blood level of free morphine over 10 ng/mL in drivers may be considered as driving under influence. Moeller et al. studied blood and urine morphine levels after consumption of poppy seed products. Five volunteers ate different kinds of poppy seed rolls or poppy seed cakes, but all blood specimens analyzed postingestion were negative for free morphine by both immunoassays and GC/MS (values below 10 ng/mL). However, GC/MS analysis of urine showed both morphine (range: 147 to 1300 ng/mL) and codeine (11–36 ng/mL). The authors conclude that eating poppy seed–containing food should not result in positive blood test for free morphine [26].

THEBAINE: A MARKER OF POPPY SEED CONSUMPTION

Thebaine is a natural constituent of poppy seed. Therefore, positive thebaine findings can be attributed to consuming poppy seed–containing food. Moreover, thebaine itself is not found in illicit heroin because it is converted primarily to thebaol and acetylthebaol during preparation of illicit heroin under the action of acetic anhydride and thus cannot be detected in urine after street heroin administration. In addition, thebaine is not available as pharmaceutical products, and it is also not a metabolite of heroin, morphine, codeine, or ethylmorphine. As a result, the presence of thebaine confirms consumption of poppy seed products. However, thebaine may not always be detected in all subjects after eating poppy seed—containing foods. In addition, concentration of thebaine is very low compared to morphine and codeine levels after eating poppy seed—containing foods [10]. In one study, thebaine was detected in concentrations ranging from 2 to 91 ng/mL in volunteers after consumption of 11 g of poppy seed [46]. Chemical structure of thebaine is given in Fig. 6.1.

Poppy Seed Defense: Does It Work at Court?

In general, most employers follow SAMHSA guideline of 2000 ng/mL cutoff for opiate testing, and it is very unlikely that after poppy seed consumption, a person would test positive at 2000 ng/mL cutoff. However, some employer still uses 300 ng/mL cutoff for opiate screening, and if a subject consumes poppy seed—containing food 3—8 h prior to drug testing, a positive opiate assay may be likely. Moreover, both morphine and codeine should be detected in urine. In addition, some herbal tea may contain opium, and drinking such tea will cause positive opiate test result both during screening and confirmation. Such test results are considered analytical positive.

For workplace drug testing, a medical review officer has the responsibility to determine whether there is any alternative explanation of analytically confirmed positive drug test. Usually if morphine concentration exceeds 2000 ng/mL and morphine/codeine ratio is > 11, such findings are considered as inconsistent with poppy seed defense. Alternatively, only presence of morphine and codeine with combined concentration less than 1000 ng/mL where the employer utilized 300 ng/mL cutoff for opiate screening may be due to

consumption of poppy seed—containing food. However, confirmation of 6-MAM (also known as 6-AM) is inconsistent with eating poppy seed—containing food because 6-MAM is a biomarker for heroin abuse. In the opinion of this author, poppy seed defense is widely used by individuals to explain positive opiate test results, but in most cases, it is ineffective because either morphine is present at significantly high levels (often exceeding 4000 ng/mL) and or 6-MAM is also confirmed in the urine.

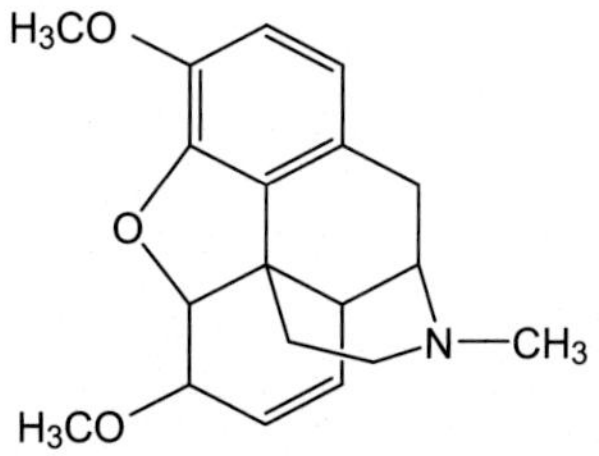

CONCLUSIONS

In order to pass workplace drug testing, this author recommends not eating any poppy seed—containing foods or drink any herbal tea at least a week prior to workplace drug testing. This is important because some employers (private employers) still use 300 ng/mL cutoff for opiate screening. Based on the expert witness experience of this author, it is hard to defend a positive drug test result conducted by a SAMHSA-certified or CAP (College of American Pathologists) forensic—certified laboratory. Therefore, best way to avoid positive drug test result is to avoid herbal tea and poppy seed—containing food. Positive cocaine test result (confirmed as benzoylecgonine, a cocaine metabolite) after consumption of Health Inca tea has been reported because such tea is made from coca leaves that still contains residual cocaine [47]. Coca tea is derived from the same plant that is commonly consumed in South America and easily obtained in the United States. Consuming such tea will also cause analytical positive test result with benzoylecgonine in workplace drug testing [48].

REFERENCES

[1] Ammone S, Hofmann U, Griese EU, Gugler N, et al. Pharmacokinetics of dihydrocodeine and its active metabolite after single and multiple oral dosing. Br J Clin Pharmacol 1999;48:317—22.

[2] Jenkins AJ, Oyler JM, Cone EJ. Comparison of heroin and cocaine concentrations in saliva with concentrations in blood and plasma. J Anal Toxicol 1995;19:359—74.

[3] Presley L, Lehrer M, Seiter W, Hahn D, et al. High prevalence of 6-acetylmorphine in morphine-positive oral fluid specimens. Forensic Sci Int 2003;133:22—5.

[4] Borriello R, Carfora A, Cassandro P, Petrella R. Clinical and forensic diagnosis of very recent heroin intake by 6-acetylmorphine immunoassay test and LC-MS/MS analysis in urine and blood. Ann Clin Lab Sci 2015;45:414—8.

[5] von Euler M, Villen T, Svensson JO, Stahle L. Interpretation of the presence of 6-monoacetylmorphine in the absence of morphine-3-glucuronide in urine samples: evidence of heroin abuse. Ther Drug Monit 2003;25:645—8.

FIG. 6.1 Chemical structure of thebaine.

[6] Moriya F, Chan KM, Hashimoto Y. Concentrations of morphine and codeine in urine of heroin abusers. Leg Med 1999;1:140—4.

[7] Konstantinova SV, Normann PT, Arnestad M, Karinen R, et al. Morphine to codeine concentration ratio in blood and urine as a marker of illicit heroin use in forensic autopsy samples. Forensic Sci Int 2012;217:216—21.

[8] Ceder G, Jones AW. Concentration ratios of morphine to codeine in blood of impaired drivers as evidence of heroin use and not medication with codeine. Clin Chem 2001;47:1980—4.

[9] Ellis AD, McGwin G, Davis GG, Dye DW. Identifying cases of heroin toxicity where 6-acetylmorphine (6-AM) is not detected by toxicological analyses. Forensic Sci Med Pathol 2016;12:243—7.

[10] Mass A, Madea B, Hess C. Confirmation of recent heroin abuse: accepting the challenge. Drug Test Anal 2018;10:54—71.

[11] Gambaro V, Argo A, Cippitelli M, Dell'Acqua L, et al. Unexpected variation of the codeine/morphine ratio following fatal heroin overdose. J Anal Toxicol 2014;38:289—94.

[12] Seif-Barghi T, Moghadam N, Kobarfard F. Morphine/codeine ratio, a key in investigating a case of doping. Asian J Sports Med 2015;6:e28798.

[13] He YJ, Brockmoller J, Schmidt H, Roots I, et al. CYP2D6 ultrarapid metabolism and morphine/codeine ratio in blood: was it codeine or heroin? J Anal Toxicol 2008;32:178—82.

[14] O'Neal CL, Poklis A. The detection of acetyl codeine and 6-acetylmorphine in opiate positive urine. Forensic Sci Int 1998;95:1—10.

[15] Brenneisen R, Hasler F, Wursch D. Acetyl codeine as a urinary marker to differentiate the use of stress heroin and pharmaceutical heroin. J Anal Toxicol 2002;26:561—6.

[16] Phillips SG, Allen KR. Acetyl codeine as a marker of illicit heroin abuse in oral fluid samples. J Anal Toxicol 2006;30:370—4.

[17] Dinis-Oliveira RJ. Metabolism and metabolomics of opiates: a long way of forensic implications to unravel. J Forensic Leg Med 2019;61:128—40.

[18] Chan P, Braithwaite RA, George C, Hylands PJ, et al. The poppy seed defense: a novel solution. Drug Test Anal 2014;6:194—201.

[19] Maas A, Krämer M, Sydow K, Chen PS, et al. Urinary excretion study following consumption of various poppy seed products and investigation of the new potential street heroin marker ATM4G. Drug Test Anal 2017;9:470—8.

[20] Sproll C, Perz RC, Lachenmeier DW. Optimized LC/MS/MS analysis of morphine and codeine in poppy seed and evaluation of their fate during food processing as a basis for risk analysis. J Agric Food Chem 2006;54(15):5292—8.

[21] Pelders MG, Ros JJ. Poppy seeds: differences in morphine and codeine content and variation in inter- and intra-individual excretion. J Forensic Sci 1996;41:209—12.

[22] Lachenmeier DW, Sproll C, Musshoff F. Poppy seed foods and opiate drug testing-where are we today? Ther Drug Monit 2010;32:11—8.

[23] Trafkowski J, Madea B, Musshoff F. The significance of putative urinary markers of illicit heroin use after consumption of poppy seed products. Ther Drug Monit 2006;28:552—8.

[24] Hill V, Cairns T, Cheng CC, Schaffewr M. Multiple aspects of hair analysis for opiates: methodology, clinical and workplace population, codeine and poppy seed ingestion. J Anal Toxicol 2005;29:696—703.

[25] Thevis M, Opfermann G, Schanzer W. Urinary concentrations of morphine and codeine after consumption of poppy seeds. J Anal Toxicol 2003;27:53—6.

[26] Moeller MR, Hammer K, Engel O. Poppy seed consumption and toxicological analysis of blood and urine samples. Forensic Sci Int 2004;143:183—6.

[27] Zentai A, Szeitzne-Szabo SM, Szabo IJ, Ambrus A. Exposure of consumers to morphine from poppy seeds in Hungary. Food Addit Contam 2012;29:403—14.

[28] Liu RH. Important considerations in the interpretation of forensic urine drug test results. Forensic Sci Rev 1992;4:51—65.

[29] Unnithan S, Strang J. Poppy tea dependance. Br J Psychiatry 1993;163:813—4.

[30] Braye K, Harwood T, Inder R, Beasley R, Robinson G. Poppy seed tea and opiate abuse in New Zealand. Drug Alcohol Rev 2007;26:215—9.

[31] Nanjayya SB, Murthy P, Chand PK, Kandaswamy A, et al. A case of poppy tea dependance in an octogenarian lady. Drug Alcohol Rev 2010;29:216—8.

[32] Seyani C, Green P, Daniel L, Pegden A. An interesting case of opium tea toxicity. BMJ Case Rep 2017;2017.

[33] King M, McDonough MA, Drummer OH, Berkovic SF. Poppy tea and the baker's first seizure. Lancet 1997;350:716.

[34] Van Thuyne W, Van Eenoo P, Delbeke FT. Urinary concentrations of morphine after the administration of herbal tea containing Papaveris fructus in relation to doping analysis. J Chromatogr B Analyt Technol Biomed Life Sci 2003;785:245—51.

[35] Powers D, Erickson S, Swortwood MJ. Quantification of morphine, codeine, and thebaine in home-brewed poppy seed tea by LC-MS/MS. J Forensic Sci 2018;63:1229—35.

[36] Steentoff A, Kaa E, Worm K. Fatal intoxication in Denmark following intake of morphine from opium poppies. Z Rechtsmed 1988;101:197—204.

[37] Haber I, Pergolizzi Jr J, LeQuang JA. Poppy seed tea: a short review and case study. Pain Ther 2019;8:151—5.

[38] Özbunar E, Aydoğdu M, Döğer R, Bostancı Hİ, et al. Morphine concentrations in human urine following poppy seed paste consumption. Forensic Sci Int 2019;295:121—7.

[39] Schuppener LM, Corliss RF. Death due to complications of bowel obstruction following raw poppy seed ingestion. J Forensic Sci 2018;63:614—8.

[40] Sproll C, Perz RC, Buschmann R. Guidelines for reduction of morphine in poppy seeds intended for food purposes. Eur Food Res Technol 2007;226:307—10.

[41] Lo DS, Chua TH. Poppy seeds: implications of consumption. Med Sci Law 1992;32:296—302.

[42] ElSohly HN, ElSohly MA, Stanford DF. Poppy seed ingestion and opiates urinalysis: a closer look. J Anal Toxicol 1990;14:308–10.

[43] Rohrig TP, Moore C. The determination of morphine in urine and oral fluid following ingestion of poppy seeds. J Anal Toxicol 2003;27:449–52.

[44] Samano KL, Clouette RE, Rowland BJ, Sample RH. Concentrations of morphine and codeine in paired oral fluid and urine specimens following ingestion of a poppy seed roll and raw poppy seeds. J Anal Toxicol 2015;39: 655–61.

[45] Meneely KD. Poppy seed ingestion: the oregon perspective. J Forensic Sci 1992;37:1158–62.

[46] Cassella G, Wu AH, Shaw BR, Hill DW. The analysis of thebaine in urine for the detection of poppy seed consumption. J Anal Toxicol 1997;21:376–83.

[47] Jackson GF, Saady JJ, Poklis A. Urinary excretion of benzoylecgonine following ingestion of health inca tea. Forensic Sci Int 1991;49:57–64.

[48] Mazor SS, Mycyk MB, Wills BK, Brace LD, et al. Coca tea consumption causes positive urine cocaine assay. Eur J Emerg Med 2006;13:340–1.

CHAPTER 7

Drug Testing in Pain Management

INTRODUCTION

Drug testing may be either for medical or for legal purposes. For medical purposes, drug testing may be conducted in a patient presented to the emergency room where a physician is suspecting a drug overdose. For such patients, a urine drug screen is ordered where FDA-approved immunoassays for 8–10 common drugs are used for detecting the presence of a drug or drug class. Such medical drug testing can also be ordered during the routine medical visit if the clinician is suspecting that the patient may be abusing nonprescription or illicit drugs. However, medical drug testing results may not have any negative consequences because the results of such testing cannot be disclosed to anyone without permission from the patient. Results obtained during medical drug testing cannot be used in any legal context because such results are also protected by patient's privacy act such as HIPPA (Health Insurance Portability and Accountability Act) regulations. As a result, a physician may choose not to order a confirmatory test for immunoassay positive test results. However, it is important to note that immunoassay positive results are considered as presumptive positive and require further validation by an alternative technique such as gas chromatography combined with mass spectrometry (GC/MS), liquid chromatography combined with mass spectrometry (LC–MS) or liquid chromatography combined with tandem mass spectrometry (LC–MS/MS) to confirm the presence of the drug or drugs in the specimen.

Legal drug testing was initiated by President Reagan who issued executive order number 12,564 on September 15, 1986. This executive order directed drug testing for all federal employees who are involved in law enforcement, national security, protection of life and property, public health, and other services requiring a high degree of public trust. Following this executive order, the National Institute of Drug of Abuse was given the responsibility of developing guidelines for federal drug testing. Currently, Substance Abuse and Mental Health Services Administration (SAMHSA) affiliated with Department of Health and Human Services of the Federal Government is responsible for providing mandatory guidelines for federal workplace drug testing. Bush et al. summarized guidelines for legal drug testing [1]. Preemployment or workplace drug testing conducted by private employers usually follow SAMHSA mandated guidelines for federal workplace drug testing. However, private employers may also test for additional drugs that are not mandated by SAMHSA guidelines, for example, benzodiazepines.

In medical drug testing, informed consent may not be taken from a patient. An overdosed patient admitted to the emergency department may not be able to grant an informed consent anyway. In contrast, in the workplace or any other legal drug testing program, obtaining informed consent before testing is mandatory. Another major difference between medical and legal drug testing is that in medical testing an initial positive screening result obtained by using immunoassays may not be confirmed by a different analytical method but for legal drug testing confirmation of presumptive positive drug during immunoassay screening is mandatory by using GC/MS or LC–MS/MS. In addition, a chain of custody must be maintained in legal drug testing indicating the name of all personnel that has possession of the specimen from the time of collection to the time of reporting results. Chain of custody is not usually initiated in medical drug testing. Therefore, a medical drug testing result may not be able to stand a legal challenge. However, there are also similarities between medical and legal drug testings because in both programs initial screening is conducted by FDA-approved immunoassays using similar cut-off concentrations. Moreover, medical drug testing like legal drug testing also screen (and confirm for legal drug testing) for commonly abused drugs. As a result, less commonly abused drugs and many designer drugs may not be detected in both medical and legal drug testing.

Drug testing in pain management is mostly focused on detecting opioids and related drugs used in pain management as well as commonly abused illicit drugs using urine specimens (rarely using oral fluid or other specimens). A major goal is to ensure patient compliance with medications but also to make sure that patients are not misusing opioids or other illicit drugs.

Fighting the Opioid Epidemic. https://doi.org/10.1016/B978-0-12-820075-9.00007-7

COMMONLY TESTED DRUGS

Commonly tested drugs include both SAMHSA mandated and non-SAMHSA drugs. For drug testing involving federal employees only SAMHSA, mandated drugs are tested. Originally, SAMHSA mandated drug testing of five commonly abused drugs including amphetamine, cocaine (tested as benzoylecgonine, the inactive metabolite), opiates, phencyclidine (PCP), and marijuana tested as 11-nor-9-carboxy Δ^9-tetrahydrocannabinol (THC-COOH), the inactive metabolite of marijuana. In the 2015 revision to proposed guidelines, SAMHSA recommended additional testing for oxycodone, oxymorphone, hydrocodone, and hydromorphone [2]. Moreover, heroin metabolite 6-monoacetylmorphine is also tested under SAMHSA protocol at a screening and confirmation cut-off of 10 ng/mL. Although phencyclidine is still included in the list of drugs mandated by SAMHSA, this drug is currently infrequently abused. Some private employers may test for additional drugs in their workplace drug testing protocols and such a comprehensive drug panel may include barbiturates, benzodiazepines, and methadone.

Testing of SAMHSA-mandated drugs in urine are summarized in Table 7.1. The testing of non-SAMHSA drugs is listed in Table 7.2.

Although drug testing can be conducted also in blood, oral fluid, sweat, hair, nail, and meconium (in newborn babies), urine is the most commonly used specimen for drug testing. For drug testing in urine, depending on the particular drug, either the parent drug or its metabolite is targeted. For SAMHSA mandated drugs, recommended cut-off concentrations for both immunoassay and GC/MS of various drugs are available. In general, such guidelines are also followed in medical drug testing. Usually, a drug or its metabolites can only be detected in urine for a limited time after the last abuse. However, detection time varies depending on the dosage administered as well as the characteristics of screening and confirmation assay. For example, marijuana (Δ^9-tetrahydrocannabinol; THC) is tested as marijuana metabolite (THC-COOH) is detectable in urine for 2−4 days (using 50 ng/mL cut-off) but for more frequent use, it may be detected up to 1 month.

TABLE 7.1
Screening, Confirmation Cut-Off Concentrations and Window of Detection in Urine for SAMHSA Drugs.

Drug Class/Drug	Target Analyte in Urine for Immunoassays	Window of Detection in Urine	Screening Cut-Off	Confirmation Cut-Off
Amphetamine/methamphetamine[a]	Amphetamine or methamphetamine	2 days	500 ng/mL	250 ng/mL
MDMA/MDA/MDEA	MDMA	2 days	500 ng/mL	250 ng/mL for all drugs
Cocaine	Benzoylecgonine	2−3 days	150 ng/mL	100 ng/mL
Opiates	Morphine	3 days for both morphine and codeine	2000 ng/mL[b]	2000 ng/mL for either morphine or codeine
Heroin	6-Monoacetyl-morphine	12 h	10 ng/mL	10 ng/mL
Hydrocodone/hydromorphone[d]	Hydrocodone	Hydrocodone or hydromorphone up to 3 days	300 ng/mL	100 ng/mL for either hydrocodone or hydromorphone
Oxycodone/oxymorphone[d]	Oxycodone	Oxycodone or oxymorphone up to 3 days	100 ng/mL	50 ng/mL for either oxycodone or oxymorphone
Marijuana	THC-COOH[c]	2−3 days after single use, up to 30 days in chronic users	50 ng/mL	15 ng/mL
Phencyclidine	Phencyclidine	8 days	25 ng/mL	25 ng/mL

MDA, 3, 4-methylenedioxyamphetamine; *MDEA*, 3, 4-methylenedioxyethylamphetamine; *MDMA*: 3, 4-methylenedioxymethamphetamine
[a] If methamphetamine is confirmed, then its metabolite amphetamine must be present at a concentration of 100 ng/mL or higher.
[b] Private employer may use 300 ng/mL cut-off for screening and confirmation although SAMHSA guideline recommends 2000 ng/mL cut-off to avoid false-positive results due to consuming poppy seed-containing foods.
[c] THC-COOH, 11-nor-Δ9-tetrahydrocannabinol- 9-carboxylic acid.
[d] From new SAMHSA proposed guidelines (2015).

TABLE 7.2
Screening and Confirmation Cut-Off Concentrations of Commonly Monitored Non-SAMHSA Drugs.

Drug Class/ Drug	Target Analyte in Urine for Immunoassays	Window of Detection in Urine	Screening Cut-Off	Confirmation Cut-Off
Barbiturates	Secobarbital	Short acting (Pentobarbital, Secobarbital, etc.): 3 days Long acting (Phenobarbital): 15 days	200/300 ng/mL	200 ng/mL
Benzodiazepines	Oxazepam or Nor-diazepam	Short acting (e.g., triazolam): 2 days Intermediate acting (e.g., lorazepam, temazepam): 5 days Long acting (e.g., diazepam): 10 days	200 ng/mL	200 ng/mL
Methadone	Methadone or EDDP (metabolite)	3 days	300 ng/mL	100 ng/mL

Drug Testing: Analytical Issues

Immunoassays are widely used as the first step of drug screening in both medical and workplace drug testing programs. Immunoassays can be easily automated, and several drugs can be analyzed using one urine specimen and results can be directly downloaded in the laboratory information system. In general, competitive immunoassays are used for drugs of abuse testing. Various commercially available immunoassays are available for drug testings including FPIA (fluorescence polarization immunoassays), EMIT (enzyme multiplied immunoassay technique), CEDIA (cloned enzyme donor immunoassay), TIA (turbidimetric immunoassay), and KIMS (kinetic interaction of microparticle in solution. The ONLINE Drugs of Abuse Testings immunoassays marketed by Roche Diagnostics (Indianapolis, IN) are based on the KIMS format. In addition, ELISA (enzyme-linked immunosorbent assay) assays are also available for certain drugs.

Usually, multiple calibrators (4–6 levels) are recommended for accurate measurements of the analyte across the entire assay range in an immunoassay although two-point calibrations are also used. Most automated assay systems can store a calibration curve depending on the assay stability of that system. Therefore, when a sample is analyzed during the period denoted by calibration stability, the assay signal is automatically converted into analyte concentration via the stored calibration curve. Drugs of abuse assays more often report "qualitative" results, that is, positive or negative with respect to a certain analyte concentration (the "cut-off" level). However, some of the assays come

in both qualitative and semiquantitative formats and in most cases such formats are defined by assay protocol and calibration. In qualitative formats, the calibration can be simplified to two calibrators, centering on the cut-off point thus providing the most accuracy around that point. The algorithm compares the signal observed with a sample with that of the cutoff calibrator and reports the result as positive or negative. Semiquantitative results can be reported using a calibration curve containing at least three calibrators (often the combination of the zero calibrators, together with two or more calibrators). However, semiquantitative results may not compare well with values obtained by more sophisticated analytic methods such as GC/MS or LC–MS/MS.

In general, immunoassays used for the drug of abuse testing target a specific drug or a metabolite. The major limitation of immunoassays is that an antibody may cross-react with a structurally similar drug causing false-positive test results. Therefore, initial drug screening should be confirmed by GC/MS, LC–MS, or LC–MS/MS. Eichhorst proposed the use of liquid chromatography combined with mass spectrometry for rapid screening and confirmation of the presence of abused drugs in urine specimens as an alternative to immunoassay screening [3].

Issues of Interferences in Immunoassay Screening Tests

Although immunoassays are used in the screening step of drugs of abuse testing, false-positive test results may occur due to the interferences of drugs different from the targeted drug. This is the reason immunoassay results should not be considered as definite proof that the target drug is present in the specimen. As a result, an alternative analytic method such as GC/MS, LC/MS, or LC–MS/MS must be used to confirm immunoassay positive drug. In legal drug testing, the confirmation step is mandatory.

Amphetamine immunoassays are subjected to interference from a long list of drugs. Several over-the-counter cold medications containing ephedrine or pseudoephedrine may cause false-positive amphetamine test results due to cross-reactivity with assay antibody. Ranitidine, an H_2 receptor blocking agent, is available over the counter without any prescription. Dietzen et al. reported that ranitidine, if present in urine at a concentration over 43 μg/mL, may cause a false-positive amphetamine screen test result using Beckman SYNCHRON immunoassay reagents (Beckman Diagnostics, Brea, CA). This concentration of ranitidine is expected in patients taking ranitidine at the recommended therapeutic dosage [4]. Casey et al. reported that bupropion, a monocyclic antidepressant and aid for smoking cessation, may cause false-positive screen results using the EMIT II amphetamine immunoassay [5]. Trazodone, an antidepressant, interferes with both amphetamine and MDMA assay. Baron et al. demonstrated that the trazodone metabolite meta-chlorophenylpiperazine is responsible for the interference [6]. Labetalol, a β-blocker commonly used for control of hypertension in pregnancy, may cause false-positive amphetamine screen results using an immunoassay. A labetalol metabolite that is structurally similar to amphetamine and methamphetamine is responsible for the interference [7]. The antidepressant desipramine and the antiviral agent amantadine also interfere with amphetamine immunoassays [8]. In addition, dimethylamylamine (DMAA), aliphatic amine naturally found in geranium flowers but also used in bodybuilding natural supplements such as Jack3d and OxyELITE Pro may cause false-positive test results with a KIMS amphetamine assay (Roche Diagnostics) and the EMIT II Plus amphetamine assay if present at a concentration of 6900 ng/mL due to structural similarity of DMAA with amphetamine [9]. Common drugs that may cause false positive with amphetamine immunoassays are listed in Table 7.3.

The antibody in the opiate immunoassay targets morphine but can detect codeine. However, many opioids such as oxycodone, oxymorphone, methadone, fentanyl, meperidine, and propoxyphene show poor cross-reactivity with various opiate assays. Therefore, specific immunoassays must be used. Opiate immunoassays like other immunoassays also suffer from providing false-positive test results due to the presence of various cross-reacting substances other than opioids

TABLE 7.3
Interferences in Immunoassays for Drug of Abuse Screening in Urine.

Immunoassay	Interfering Drugs
Amphetamine/methamphetamine	Ephedrine, pseudoephedrine and phenylephrine are most commonly encountered but other drugs including ranitidine, bupropion, trazodone, labetalol, desipramine, amantadine, chlorpromazine, and chloroquine
Opiate	Levofloxacin, ofloxacin, pefloxacin, gatifloxacin, moxifloxacin, ciprofloxacin, and norfloxacin
Methadone	Quetiapine, diphenhydramine, doxylamine, and verapamil metabolites
Marijuana	Niflumic acid, efavirenz, ibuprofen, and naproxen
Phencyclidine	Dextromethorphan, ibuprofen, thioridazine, and venlafaxine
Benzodiazepine	Oxaprozin and sertraline
Barbiturates	Ibuprofen and naproxen

in urine specimens. Certain quinolone antibiotics may cause false-positive test results with opiate immunoassay screening. Baden et al evaluated potential interference of 13 commonly used quinolones (levofloxacin, ofloxacin, pefloxacin, enoxacin, moxifloxacin, gatifloxacin, trovafloxacin, sparfloxacin, lomefloxacin, ciprofloxacin, clinafloxacin, norfloxacin, and nalidixic acid) with various opiate immunoassays (at 300 ng/mL cut-off concentration) and observed that levofloxacin and ofloxacin may cause false-positive opiate test results with assays manufactured by Abbot Laboratories for application on the AxSYM analyzer (Abbott Laboratories, Abbott Park, IL). In addition, such interferences were also observed with CEDIA, EMIT II, and Abuscreen ONLINE assay (Roche Diagnostics, Indianapolis, IN). Moreover, pefloxacin administration may cause false positives with CEDIA, EMIT II, and Abuscreen ONLINE assay, gatifloxacin with CEDIA and EMIT II assays, lomefloxacin, moxifloxacin, ciprofloxacin, and norfloxacin with the Abuscreen ONLINE assay [10]. Rifampicin is used in treating tuberculosis and may cause false-positive test results with opiate immunoassays such as the KIMS (kinetic interaction of microparticle in solution methods) assay on the Cobas Integra analyzer (Roche Diagnostics, Indianapolis, IN). A false-positive result may be observed even after 18 h of administration of a single oral dose of 600 mg of rifampicin [11].

False-positive test results with methadone immunoassays in three schizophrenia patients treated with quetiapine monotherapy have been reported [12]. Rogers et al. reported positive methadone urine drug test results in a patient using the One Step Multi-Drug, Multi-Line Screen Test Devices (ACON Laboratories, San Diego, CA), a point of care device for a urine drug screen. The patient had no history of methadone exposure but ingested diphenhydramine [13]. Doxylamine intoxication may cause false-positive results with both EMIT d.a.u opiate and methadone assays. The urine doxylamine concentration needed to cause positive test results was 50 μg/mL for methadone and 800 μg/mL for opiate [14]. Verapamil metabolites also interfere with methadone assay for the application on Olympus AU5000 analyzer [15].

Although uncommon, false-positive marijuana test results may occur during the screening step due to cross-reactivity from other compounds that are not illicit drugs. Boucher et al. described a case of a 3-year-old girl who was hospitalized because of behavioral disturbance of unknown cause. The only remarkable finding in her medication history was the suppositories of niflumic acid that was initiated 5 days before hospitalization. After admission, her urinary toxicology screen was positive for the presence of marijuana metabolite but parents strongly denied such exposure. Further analysis of the specimen using chromatography failed to confirm the presence of marijuana metabolite but niflumic acid was detected in the specimen. The authors concluded that the false-positive marijuana test result was due to the presence of niflumic acid in the urine specimen [16]. The antiviral agent efavirenz is known to cross-react with marijuana immunoassays [17]. Nonsteroidal drugs naproxen and ibuprofen may also cause false-positive test results with marijuana immunoassay [18].

Dextromethorphan is an antitussive agent that is found in many over the counter cough and cold medications. Ingesting high amounts of dextromethorphan (over 30 mg) may result in positive false-positive test results with opiate and PCP immunoassays [19]. Thioridazine is known to cause false-positive PCP test with both EMIT d.a.u and EMIT II phencyclidine immunoassays [20]. Venlafaxine and its metabolites also may cause false-positive PCP test with immunoassay [21].

Oxaprozin, a nonsteroidal antiinflammatory drug, may cause false-positive test results with CEDIA, EMIT, and FPIA benzodiazepine immunoassays using a cut-off of 200 ng/mL [22]. Sertraline may also cause a false-positive test result with benzodiazepine immunoassay [23].

Barbiturate immunoassays can recognize a wide variety of barbiturates including amobarbital, butalbital, pentobarbital, secobarbital, and phenobarbital. Secobarbital is used as a calibrator in several commercially available immunoassays for barbiturates. Acute ingestion of ibuprofen and chronic ingestion of naproxen may cause false-positive test result with barbiturate immunoassay [18].

Drug Confirmation Using Mass Spectrometry

Drugs tested positive during immunoassay screening must be confirmed by an alternative technique to ensure the drug and or metabolite was indeed present in the urine specimen. GC/MS is widely used for drug confirmation but certain drug and/or metabolite requires derivatization step to convert a polar molecule to a nonpolar molecule for GC/MS analysis because polar molecules cannot be analyzed by this method. For example, for the analysis of amphetamine and methamphetamine, these drugs can be extracted from urine using solid-phase extraction or liquid–liquid extraction. However, derivatization is needed before GC/MS analysis. For nonchiral derivatization, commonly used agents include pentafluoropropionic anhydride, heptafluorobutyric anhydride, trifluoroacetic anhydride, 4-carboethoxyhexafluorobutyryl chloride (4-CB), N-methyl-N-t-butyldimethylsilyl trifluoroacetamide, N-trifluoroacetyl-1-propyl chloride, and 2,2,2-trichloroethyl chloroformate. However, pentafluoropropionic anhydride and heptafluorobutyric anhydride are commonly used. Deuterated amphetamine and methamphetamine can be used as internal standards. MDMA and MDA can be confirmed by GC/MS using similar extraction and derivatization protocol.

More recently LC–MS or LC–MS/MS are gaining acceptance as a preferred methodology for drug confirmation. One advantage is that no derivatization step is necessary because liquid chromatography is capable of analyzing both polar and nonpolar compounds. In addition, multiple drugs can be analyzed in one step after minimal sample preparation. For systematic toxicology analysis LC–MS/MS is the best approach where mass spectrometer is operated in the selected ion monitoring mode or multiple reaction monitoring modes because drugs can be easily identified from product ion spectra library associated with the method. More recently dilute and shoot approach has been introduced for the analysis of many drugs in one run. However, possibilities of matrix effect and false-positive results are associated with such an approach when no extraction is conducted. More recently, high-resolution mass spectrometry has been used in toxicology laboratories. High-resolution mass spectrometry allows determining the exact molecular weight of the compound so that a potential drug or poison could be identified from a list of accurate masses of several potential drugs, poisons, or pesticides. However, it is important to remember that several drugs or metabolites may have the same empirical formula and molecular mass. Therefore, care must be taken in interpreting results [24]. Tsai et al. used an ultrahigh-performance liquid chromatography—quadrupole time-of-flight mass spectrometry for the screening and confirmation of 62 drugs of abuse and their metabolites in urine. These drugs include not only the most commonly abused drugs such as amphetamines, opioids, cocaine, benzodiazepines, and barbiturates but also other new and emerging abused drugs. Urine samples were diluted fivefold with deionized water before analysis. Using a superficially porous microparticulate column and an acetic acid-based mobile phase, 54 basic and 8 acidic analytes could be detected within 15 and 12 min in positive and negative ionization modes, respectively [25]. In another report, the authors developed a rapid, reproducible, and sensitive reversed-phase liquid chromatography—mass spectrometry method for the identification and semiquantitative confirmation of stimulants in urine. The method was capable of separating compounds such as cocaine and its metabolites, amphetamines, substituted cathinone, and other designer drugs, with a total run time of 11 min. The method was subsequently used to confirm the presence of these stimulants in the urine of patients attending the Drug Treatment Center [26].

DRUG TESTING IN PAIN MANAGEMENT

The opioid overdose crisis is a major public health concern and as a result, the US government has declared the opioid crisis as a national public health emergency. The Joint Commission released new public standards for pain management in 2017 that emphasized safe opioid prescription policy. The US Surgeon General issued an unprecedented letter to physicians in 2016 calling for ways to end the opioid crisis [27]. Prescription opioids alone are responsible for 46 overdose deaths every day and over 40% of all opioid-related overdose deaths [28]. Therefore, it is a difficult balancing act for physicians to protect patients who merit access to opioids for controlling both cancer and noncancer-related pain and also to employ an effective strategy to reduce nonmedical opioid use. To achieve this goal, physicians must perform an initial screen before prescribing opioids to determine the risk of nonprescription opioid use by a patient. In addition, implementing effective strategies to prevent nonmedical opioid use is also important for physicians who prescribe opioid pain medications. Urine drug testing in pain management is an important strategy to prevent nonmedical opioid use. CDC recommendations #10; "Guideline for Prescribing Opioids for Chronic Pain States" states that when prescribing opioids for chronic pain clinicians should use urine drug test testing before starting opioid therapy and consider urine drug testing at least annually to assess for prescription medication as well as other controlled prescription drugs and illicit drugs. Requiring a urine drug test does not imply a lack of trust on the part of the provider but a standardized set of safety measures offer to all patients taking opioids for pain control. Clinicians must discuss the purpose of urine drug tests with the patient and also should discuss potential cost if the insurance is not paying for such testing. Clinicians should also discuss with patients results of urine drug testing covering both expected and unexpected findings (if any). However, it is important to interpret results very carefully and a patient should not be dismissed from the clinical solely based on unexpected urine drug testing results [29].

Argoff and his colleague based on a review of several guidelines for pain management drug testing published a consensus guideline. Definitive (e.g., chromatography-based) testing is recommended as most clinically appropriate for urine drug testing because of its accuracy; however, institutional or payer policies may require initial use of presumptive testing (i.e., immunoassay). The rational choice of substances to analyze for drug panel involves considerations that are specific to each patient and related to illicit drug availability. Appropriate opioid risk stratification is based on patient history (especially psychiatric conditions or history of opioid or substance use disorder), prescription drug monitoring program data, results from validated risk assessment tools, and previous urine drug testing results. Urine drug monitoring is suggested to be performed at baseline for most patients prescribed opioids for chronic pain and at least annually for those at low risk, two or more times per year for those at moderate risk, and three or more times per year for those at high risk. Additional urine drug testing should be performed as needed based on clinical judgment [30].

Drugs Usually Tested in Pain Management Panel

Drugs that should be considered when monitoring chronic noncancer pain patients include morphine, oxycodone, oxymorphone, hydrocodone, hydromorphone, methadone, fentanyl, and buprenorphine. However, some laboratories may include meperidine, dihydrocodeine, and propoxyphene in the list of opioids tested along with commonly prescribed benzodiazepines because benzodiazepines may also be prescribed for management along with opioids in some patients. Because a patient may also use illicit drugs pain management drug testing often includes amphetamines, benzodiazepines (although these may be prescribed therapeutically), cannabinoids, cocaine (as benzoylecgonine, the inactive metabolite), and PCP. Although abuse of barbiturates is low, some laboratories may also screen for the presence of barbiturates in urine using immunoassays. The screening and confirmation cut-off in pain management drug testing may differ from SAMHSA guidelines. For example, in SAMHSA-mandated drug testing, the cut-off concentration is 2000 ng/mL for the immunoassay screening of opiates. This is due to avoid false-positive test results in people eating poppy seed-containing food. However, for pain management, the cut-off concentration for the screening of opiates in urine specimens is 300 ng/mL. Typical screening and confirmation cut-off levels of various drugs commonly included in pain management drug testing panel using urine are summarized in Table 7.4.

A variety of biological specimens can be used for drug testing including urine, blood, sweat, saliva (oral fluid), hair, and nails. However, urine is the most commonly used matrix for drug testing because the collection is noninvasive, the window of detection is longer than blood or saliva, and concentrations of drugs and metabolites also tend to be higher in the urine. Moreover, urine drug testing methodologies are well validated and very reproducible. Therefore, for pain management monitoring, urine is the most commonly used specimen. However, oral fluid is gaining acceptance as an alternative matrix for pain management drug testing. One advantage of pain management drug testing using oral fluid is that drug levels may correlate with serum levels and provide some clinical information. In contrast, drug levels monitored using urine specimens cannot be correlated with serum values.

URINE DRUG TESTING IN PAIN MANAGEMENT

Pain medicine physicians are increasingly using urine drug tests to monitor the adherence of patients to chronic opioid therapy. It is also important to know if the patient is using other drugs not prescribed to the patient or illicit substances because research indicates that some patients taking prescription opioids may abuse nonprescription opioids or illicit substances due to biological and environmental susceptibility factors. These individuals often report that they develop a craving and higher tolerance for opioids and take more medication than prescribed. Moreover, aberrant drug-related behavior sometimes associated with addiction is perpetuated by a physiological drive that comes with using opioids or discontinuing opioid therapy. Physicians can encounter problems in prescribing opioids for some patients with chronic pain such as multiple unsanctioned dose escalations, episodes of lost or stolen prescriptions, and positive urine drug screenings for illicit substances. Michna et al. explored the use of questions on abuse history in predicting problems with prescribing opioids for patients at a hospital-based pain management program using 145 patients who were taking long- and short-acting opioids for their pain. The authors classified patients as high or low risk based on their responses to interview questions about (1) substance abuse history in their family, (2) past problems with drug or alcohol abuse, and (3) history of legal problems. The treating physicians completed a questionnaire about problems that they had encountered with their patients. Problem behaviors were verified through chart review. Patients who admitted to a family history of substance abuse, a history of legal problems, and drug or alcohol abuse were prone to more aberrant drug-related behaviors, including a higher incidence of lost or stolen prescriptions and the presence of illicit substances in their urine. Patients classified as high risk also had a significantly higher frequency of reported mental health problems and motor vehicle accidents. More of these patients smoked cigarettes, tended to need a cigarette within the first hour of the day, took higher doses of opioids, and reported fewer adverse effects from the medications than did those without such a history. The authors concluded that questions about abuse history and legal problems can be useful in predicting aberrant drug-related behavior with opioid use in persons with chronic noncancer pain [31].

TABLE 7.4
Typical Pain Management Drug Panel.[a]

Drug Class	Individual Drug	Typical Immunoassay Cut-Off	Typical Confirmation Cut-Off
Opiate	Codeine	300 ng/mL (codeine + morphine)	100 ng/mL
	Morphine	300 ng/mL	100 ng/mL
	Hydrocodone	(hydrocodone + hydromorphone)	100 ng/mL
	Hydromorphone	100 ng/mL	100 ng/mL
	Oxycodone	(oxycodone + oxymorphone)	50 ng/mL
	Oxymorphone		50 ng/mL
Illicit opiate	6-Monoacetyl morphine (heroin metabolite)	10 ng/mL	10 ng/mL
Synthetic opioid	Buprenorphine	5 ng/mL (buprenorphine + nor-buprenorphine)	2 ng/mL
	Nor-buprenorphine		2 ng/mL
	Fentanyl	2 ng/mL (fentanyl + nor-fentanyl)	2 ng/mL
	Nor-fentanyl	300 ng/mL	2 ng/mL
	Methadone	300 ng/mL	100 ng/mL
	EDDP	200 ng/mL	100 ng/mL
	Tramadol	100 ng/mL	50 ng/mL
	Tapentadol		50 ng/mL
Benzodiazepine	Alprazolam	200 ng/mL for all benzodiazepines plus metabolites	40 ng/mL
	α−OH−alprazolam		20 ng/mL
	Clonazepam		20 ng/mL
	7-Aminoclonazepam		20 ng/mL
	Diazepam		50 ng/mL
	Nor-diazepam		50 ng/mL
	Lorazepam		50 ng/mL
	Oxazepam		50 ng/mL
	Temazepam		
Amphetamines	Amphetamine	500 ng/mL	100 ng/mL
	Methamphetamine	500 ng/mL	100 ng/mL
	MDMA	500 ng/mL	100 ng/mL
	MDA	500 ng/mL	100 ng/mL
Cocaine	Benzoylecgonine (metabolite)	300 ng/mL	150 ng/mL
Marijuana	THC-COOH (metabolite)	50 ng/mL	20 ng/mL
Phencyclidine	Phencyclidine	25 ng/mL	25 ng/mL
Alcohol	Alcohol (ethyl alcohol)	10 mg/dL	
	Ethyl glucuronide (metabolite)	500 ng/mL	

The drugs listed are typical drugs tested in pain management panel but some lab may perform testing of additional drugs such as propoxyphene, barbiturates and muscle relaxants (carisoprodol and meprobamate).

[a] This is an example of typical pain management drug panel but confirmation cut-offs may vary from lab to lab. In addition, creatinine in urine is tested (should be 20−400 mg/dL) along with temperature and specific gravity to ensure that specimen is not adulterated. Moreover, some lab may perform additional adulteration check.

Studies have found that pain management patients are unreliable in their report of misuse of prescription medications and the use of other illicit substances. Berndt et al. interviewed 109 consecutive patients predominantly with facial, neuropathic or back pain were about present medication at first admission to the pain clinic. Reports given by the patients were verified by toxicological urine screening, mainly with thin-layer chromatography and confirmation using gas chromatography−mass spectrometry. The authors observed that, in only 74 patients (68%), urine drug testing results corresponded with drugs taken by patients. However, 23 patients (21%) concealed the consumption of drugs, and 2 patients (2%) did not take their medications. Ten cases were not interpretable. The authors further reported that 54% of the drugs concealed were psychotropic substances, mostly benzodiazepines, and 42% were analgesic combinations, partly with psychotropic additives. Drug intake was concealed significantly more often with polypharmacy that was occurring more frequently in patients with headache or facial pain, longer duration of pain, young age, psychiatric diagnosis, and history of substance abuse. Patients with initial noncompliance were more likely to conceal drug consumption in follow-up investigations as well. Therefore, screening for medication compliance in patients with chronic nonmalignant pain is recommended, especially in those with the risk factors for abusing illicit substances [32].

Fishbain et al. investigated what percentage of chronic pain patients do not properly disclose their drug use to physicians. Urine was tested for benzodiazepines, opioids, tricyclics, propoxyphene, cannabinoids, barbiturates, amphetamines, methadone, methaqualone, phencyclidine, alcohol, and cocaine. Toxicology reports were negative in 121 (53.5%) and positive in 105 (46.5%) patients. Of the 226 patients, 8.4% had illicit drugs in the urine (6.2% cannabis, 2.2% cocaine). Twenty (8.8%) of the patients provided incorrect self-report information about current drug use, the incorrect information most frequently about illicit drugs. Drug urine toxicology sensitivity results indicated that a significant percentage of patients was claiming to be taking a drug but was not taking it or taking it incorrectly. Patients who were more likely to provide incorrect self-report information about current drug use were more likely to be younger, to be a workers' compensation, and to have been assigned a diagnosis of polysubstance abuse in remission [33].

In another study, the authors performed a retrospective analysis of data from 470 patients who had urine screening at a pain management program in an urban teaching hospital. Urine samples were analyzed using gas chromatography–mass spectrometry. Patients were categorized as having urine screens that were "normal" (expected findings based on their prescribed drugs) or abnormal. Abnormal findings were those of [1] absence of a prescribed opioid [2], presence of an additional nonprescribed controlled substance [3], detection of an illicit substance, and [4] an adulterated urine sample. Forty-five percent of the patients were found to have abnormal urine screens. Twenty percent were categorized as having an illicit substance in their urine. Illicit substances and additional drugs were found more frequently in younger patients than in older patients. The authors concluded that random

urine toxicology screens among patients prescribed opioids for pain reveal a high incidence of abnormal findings. Common patient descriptors, and number, type, and dose of prescribed opioids were found to be poor predictors of abnormal results [34].

A number of studies indicate that 10.8%–34% of patients with chronic pain use illicit drugs. One hypothesis for this occurrence is that some patients may be supplementing their prescription medications with illicit drugs. In one study using urine specimens from a cohort of nearly 400,000 patients whose identities had been redacted, and who were being treated for chronic pain with opioid therapy, the authors correlated the patients' positivity with their prescribed medication to the prevalence of illicit substance use. A secondary study was conducted to correlate the amount of prescribed medication excreted in urine (measured in ng/mL) with the incidence of illicit drug use. The specific prescription medications analyzed were hydrocodone, morphine, and oxycodone. The authors observed that specimens that were negative for prescribed hydrocodone (27.3%), morphine (11.5%), or oxycodone (19%) were more likely to contain illicit drugs than those found to be positive for the prescribed medication. The illicit drug prevalence among the inconsistent specimens was 15.3% for hydrocodone, 23.8% for morphine, and 24.4% for oxycodone. The secondary study showed no statistically significant difference in the excretion level of prescribed medication between those patients using and not using illicit drugs. This work supports the hypothesis that people who are positive for their prescribed medications use fewer illicit drugs than those who do not take their medications. It may be beneficial for physicians to test more thoroughly for illicit drugs when patients' drug tests are negative for their prescribed medications [35].

Limitations of Pain Management Drug Testing Using Immunoassays

For pain management drug testing using urine specimens, initial drug screening using immunoassays is still the most commonly used methodology due to low cost and ease of adaptation of such tests on automated chemistry analyzers. Therefore, rapid turnaround time can be achieved. A typical testing menu should cover commonly prescribed pain medications including immunoassays for oxycodone (also detects oxymorphone), hydrocodone (also detects hydromorphone), opiate (detects codeine and morphine), fentanyl, buprenorphine, methadone, as well as immunoassays for benzodiazepines, 6-monoacetylmorphine, amphetamines, cocaine (targets its inactive metabolite benzoylecgonine), marijuana (as THC-COOH, the inactive metabolite), and phencyclidine. Most of these tests are also available as a point of care test that may be waived test and such tests can be conducted in the physician's office offering the advantage of rapid turnaround time and allowing a provider to discuss results with the patient. If the urine specimen is sent to a laboratory, turnaround time will be certainly longer.

Immunoassays have many limitations. Most immunoassays are qualitative assay where results above a predetermined cut-off are reported as positive. However, some immunoassays could be operated in a semiquantitative way but, semiquantitative values may not reflect drug level determined by an LC–MS/MS method. In one study using 48 urine specimens collected from pain management patients taking oxycodone, the authors reported that semiquantitative values obtained by the DRI oxycodone immunoassay did not match with true oxycodone and metabolite oxymorphone concentrations combined obtained by using LC/MS/MS [36]. Similarly, semiquantitative values produced by DRI hydrocodone/hydromorphone assay also showed poor correlation with values determined by the LC–MS/MS method [37].

Another major limitation of immunoassays is the lack of sensitivity. For example, cut-off concentration for detecting hydromorphone using immunoassays is usually 300 ng/mL. However, some patients receiving prescription hydromorphone may not produce enough hydromorphone concentration in urine especially if receiving low dose hydromorphone. In one study, the authors reported that 69% of the specimens that were positive for hydromorphone by LC–MS/MS would have been falsely scored as negative by immunoassay [38]. In addition, concurrent marijuana abuse is associated with a reduction of urine opioid levels. In one study using a dataset of approximately 800,000 urine drug test results collected from pain management patients over a time frame of multiple years, creatinine corrected opioid levels were evaluated to determine if the presence of the primary marijuana marker 11-nor-carboxy-tetrahydrocannabinol (THC-COOH) was associated with statistical differences in excreted opioid concentrations. The authors observed that for each of the opioids investigated (codeine, morphine, hydrocodone, hydromorphone, oxycodone, oxymorphone, fentanyl, and buprenorphine), marijuana use was associated with statistically significant lower urinary opiate levels than in samples without indicators of marijuana use [39].

Another limitation of using immunoassays for urine specimens collected from pain management patients is that physicians may order a wrong test. A common

mistake is to order an opiate screen to evaluate compliance with oxycodone medication. Although oxycodone is an opioid, it has very poor cross-reactivity with opiate immunoassay where antibody recognizes morphine and codeine. A 40-year-old man receiving 20 mg oxycodone twice a day for headache, routinely called the clinic stating that he finished his medication faster and needed a refill. His urine drug test was negative, and suspecting he was selling oxycodone, he was dismissed from the clinic. A family member contacted a toxicologist who informed that oxycodone may cause false-negative urine opiate drug screen. An aliquot of the original urine specimen was retested using GC/MS and oxycodone level of 1124 ng/mL was confirmed. In this case, a complaint patient was wrongly denied oxycodone because the clinician ordered the wrong test [40].

Sometimes benzodiazepines are also prescribed to pain management patients and compliance with such medication is tested by using benzodiazepine immunoassays at 200 ng/mL cut-off. However, benzodiazepine immunoassays are inadequate for testing of compliance because antibodies used in the immunoassays have poor cross-reactivity with several benzodiazepines. West et al. investigated more appropriate cutoffs for compliance monitoring of patients prescribed clonazepam as determined using immunoassay and LC–MS/MS. The authors selected this benzodiazepine because it forms one major metabolite, 7-aminoclonazepam that is specific for that drug. Patients whose only benzodiazepine medication was clonazepam were selected as the test population and urine specimens were tested using the Microgenics DRI benzodiazepine assay with a 200 ng/mL cutoff. The same samples were quantitatively assessed for 7-aminoclonazepam by LC–MS/MS with a cutoff of 40 ng/mL. The authors analyzed specimens from 180 patients and observed positivity rates of only 21% (38 samples) by immunoassay. The positivity rate was 70% (126 samples) if the LC–MS/MS cutoff was set at 200 ng/mL. However, the positivity rate was 87% (157 samples) if the LC–MS/MS was set at 40 ng/mL. Concentration distributions revealed that 7% patients showed clonazepam and metabolite concentrations in the 40–100 ng/mL range. The authors concluded that this low immunoassay positivity rate is inconsistent with the manufacturer's published cross-reactivity data for clonazepam and 7-aminoclonazepam. These data illustrate the limitations of using a 200 ng/mL cutoff to monitor clonazepam compliance and suggest that a cutoff of 40 ng/mL or less is needed to reliably monitor the use of this drug [41].

Darragh et al. hypothesized that immunoassay-based methods lack the requisite sensitivity for detecting benzodiazepine use primarily due to their poor cross-reactivity with several major urinary benzodiazepine metabolites. A High Sensitivity Cloned Enzyme Donor Immunoassay (HS-CEDIA), in which β-glucuronidase is added to the reagent, has been shown to perform better than traditional assays. To determine the diagnostic accuracy of HS-CEDIA, as compared to the Cloned Enzyme Donor Immunoassay (CEDIA) and Kinetic Interaction of Microparticles in Solution (KIMS) screening immunoassays and LC−MS/MS, for monitoring benzodiazepine use in patients treated for chronic pain the authors analyzed a total of 299 urine specimens from patients treated with benzodiazepines using the HS-CEDIA, CEDIA, and KIMS assays. The sensitivity and specificity of the screening assays were determined using the LC−MS/MS results as the reference method.

Of the 299 urine specimens tested, 141 (47%) confirmed positive for one or more of the benzodiazepines/metabolites by LC−MS/MS. The CEDIA and KIMS sensitivities were 55% (78/141) and 47% (66/141), respectively. Despite the relatively higher sensitivity of the HS-CEDIA screening assay (78%; 110/141), primarily due to increased detection of lorazepam, it still missed 22% (31/141) of benzodiazepine-positive urines. The KIMS, CEDIA, and HS-CEDIA assays yielded accuracies of 75%, 79%, and 90%, respectively, in comparison with LC−MS/MS. The authors concluded that while the HS-CEDIA provides higher sensitivity than the KIMS and CEDIA assays, it still missed an unacceptably high percentage of benzodiazepine-positive samples from patients treated for chronic pain. LC−MS/MS quantification with enzymatic sample pretreatment offers superior sensitivity and specificity for monitoring benzodiazepines in patients treated for chronic pain [42].

Application of Mass Spectrometry for Pain Management

Due to many limitations of immunoassays, GC/MS or LC−MS/MS are considered to be a better option for determining the compliance of pain management patients with their medication. Although GC/MS methods are still used for monitoring opioids and benzodiazepines in urine specimens collected from pain management patients, more recently, LC−MS/MS-based methods are used for that purpose. With the application of mass spectrometry, individual drugs can be identified unambiguously by their mass spectral characteristics at much lower concentrations than cut-off levels used in conventional immunoassays. Therefore, mass spectrometry-based methods offer superior sensitivity and specificity compared to immunoassays.

Codeine is metabolized to morphine and then morphine is conjugated with glucuronic acid to form morphine-3-glucuronide and morphine-6-glucuronide. Other opioids are also metabolized and then conjugated. Therefore, these glucuronide metabolites must be hydrolyzed into free drugs before extraction for GC/MS analysis. Both acid hydrolysis and enzymatic hydrolysis can be used to generate free drug but enzyme hydrolysis using β-glucuronidase may require much longer incubation time unless the elevated temperature is used to accelerate enzymatic hydrolysis. Although morphine and codeine can be analyzed after derivatization to respective pentafluoropropionyl or trimethylsilyl derivative, keto-opioids such as oxycodone and oxymorphone may cause a problem during quantification. One approach to circumvent this problem is to use hydroxylamine to convert keto-opioids into oxime derivative and followed by a second derivatization step. In this approach, multiple opioids can be analyzed in one run.

Broussard et al. described a method for simultaneous identification and quantitation of codeine, morphine, hydrocodone, and hydromorphone in urine as trimethylsilyl and oxime derivatives using GC/MS. After adding deuterate codeine, morphine, hydrocodone, and hydromorphone in urine, the authors hydrolyzed conjugated metabolites already present in urine using β-glucuronidase. For achieving complete hydrolysis, the specimen was incubated at 56°C for 2 h with β-glucuronidase. After hydrolysis, aqueous hydroxylamine was added for the derivatization of keto opiates. Then derivatized hydrocodone, hydromorphone along with free morphine, codeine, and internal standards were extracted from urine using solid phase-bonded silica extraction columns (Varian, Palo Alto, CA). Further derivatization (trimethylsilyl derivatives) was performed using N−O-bis (trimethylsilyl) trifluoroacetamide and trimethylchlorosilane followed by analysis using GC/MS. The mass spectrometer was operated under selected ion monitoring mode and ions selected for codeine were m/z 243, 343, and 371 while ions selected for d3-codeine were 346 and 374. For quantitation, m/z 371 (for codeine) and 374 (for d3-codeine) were selected. Similarly, for morphine, ions selected were m/z 234, 401, and 429; whereas, for d3-morphine, ions selected were 417 and 432. In addition, m/z 429 and 432 were used for quantitation. For hydrocodone m/z 297, 371, and 386 were selected and for d3-hydrocodone m/z 300 and 389 were selected. For quantitation, m/z 386 and m/z 389 were selected. For the analysis of hydromorphone, m/z 355, 429, and 444 were chosen; whereas, for the analysis of d3-hydromorphone, m/z 358 and 447 were selected. For quantification, m/z 355 and m/z 358 were used [43].

Various opioids can also be analyzed using LC—MS/MS. Cone et al. used LC—MS/MS for the analysis of codeine, nor-codeine, morphine, hydrocodone, nor-hydrocodone, hydromorphone, dihydrocodeine, oxycodone, nor-oxycodone, and oxymorphone in urine after the hydrolysis of conjugates using β-glucuronidase [44]. A significant proportion of opioids, such as morphine, codeine, hydromorphone, oxymorphone, buprenorphine, and nor-buprenorphine, are detected as glucuronide metabolites in urine and sometimes only glucuronide metabolites are present in urine. Therefore, the hydrolysis of conjugates before analysis may provide better results in LC—MS/MS analysis of opioids. Although acid hydrolysis is efficient in converting glucuronide metabolites to free drug, it destroys 6-monoacetylmorphine (6-MAM), a unique metabolite of heroin. Therefore, various sources of β-glucuronidase, such as *Escherichia coli*, *Helix pomatia*, *Patella vulgate*, and abalone, have been utilized in opioid hydrolysis, with each exhibiting different hydrolysis efficiency and optimal conditions. Protein precipitation provides a fast and efficient approach for sample cleanup and removal of β-glucuronidase enzymes in urine following hydrolysis and improves the sensitivity and precision of the LC—MS/MS method. Therefore, the objectives of the study by Yang et al. were to conduct a comprehensive evaluation of the performance of four different β-glucuronidase enzymes and three protein precipitation plates and optimize the sample preparation workflow for a qualitative opioid confirmation assay. In addition, this work aimed to develop and validate a qualitative LC—MS/MS method to replace a GC/MS opioid confirmation assay in which sample preparation was very time-consuming and laborious. A recombinant β-glucuronidase exhibited the best overall hydrolysis efficiency for seven opioid glucuronide conjugates compared with β-glucuronidase from red abalone, *Escherichia coli*, and *Patella vulgata*. Following hydrolysis, 100 μL of the urine samples was transferred to a protein precipitation plate. Then, 300 μL of acetonitrile was added to each well and mixed with the sample thoroughly for cleanup and removal of the enzyme. Separation of 13 opioids was performed using an Agilent high-performance liquid chromatography with a Kinetex 2.6 μm Phenyl-Hexyl column heated at 30°C with gradient elution at a flow rate of 0.7 mL/min. Mobile phase A consisted of 10 mM ammonium formate, and mobile phase B consisted of 0.1% formic acid in methanol. The gradient started at 20% B was linearly increased from 20% to 40% B from 0 to 0.7 min, then increased from 40% to 100% B from 0.7 to 4.5 min, held at 100% B to 6 min, and reequilibrated at the initial condition for 2 min. The injection volume was 10 μL. A 5500 QTrap mass spectrometer was utilized with an electrospray ionization source operated in the positive mode. Multiple reaction monitoring was used with two transitions monitored for each compound and one transition for the internal standards. Retention time and ions monitored for various opioids including some metabolites for analysis by LC—MS/MS are listed in Table 7.5.

Sixty-two patient urine samples were analyzed by both the GC—MS and LC—MS/MS methods by the authors. A total of 40 morphine, 19 codeine, 13 hydrocodone, 22 hydromorphone, 26 oxycodone, and 26 oxymorphone peaks were identified by GC/MS. Of these peaks, 96.6% were confirmed by the LC—MS/MS method. The LC—MS/MS assay detected an additional three patient samples that were positive for morphine, one for codeine, six for hydromorphone, five for hydrocodone, two for oxycodone and three for oxymorphone, which was reported as either negative or inconclusive by the GC—MS assay. These additional opioids were all confirmed by patient prescription information or were supported by the presence of metabolites and/or parent drugs in the same sample. Twenty-two patient urine samples were tested positive for 6-MAM, methadone/EDDP, fentanyl/norfentanyl, and buprenorphine/norbuprenorphine using a reference LC—MS/MS method from an outside laboratory. The reference LC—MS/MS assay identified a total of five 6-MAM, nine methadone, eight EDDP, six fentanyl, six norfentanyl, five buprenorphine, and six norbuprenorphine peaks. Of those peaks, 93.3% were confirmed by authors LC—MS/MS method. Three peaks were called negative by authors LC—MS/MS assay as they were below the cut-offs: one methadone (77 ng/mL), one buprenorphine (8.7 ng/mL), and one norbuprenorphine peak (58 ng/mL). It also detected an additional buprenorphine peak (21 ng/mL). The buprenorphine peak was found in the presence of norbuprenorphine, and the reference method detected only norbuprenorphine [45].

Benzodiazepines are metabolized followed by conjugation before being excreted in the urine. Therefore, the hydrolysis of conjugate using β-glucuronidase before extraction is recommended. Benzodiazepines and their metabolites can be analyzed by GC/MS as trimethylsilyl derivatives that can be prepared by using N—O-bis (trimethylsilyl) trifluoroacetamide [46]. Instead of trimethylsilyl derivative, tert-butyldimethylsilyl derivatives can also be used for GC/MS analysis and such derivatization can be achieved by using N-methyl-N-(tert-butyldimethylsilyl)-trifluoroacetamide. West and Ritz analyzed oxazepam,

TABLE 7.5
Retention Time and Ions Monitored for Various Opioids Including Some Metabolites for Analysis by LC—MS/MS.

Opioid	Retention Time	Parent Ion (m/z)	Daughter Ion (m/z)	Linearity
Morphine	1.47 min	286.1	165.1, 152.2	25—2000 ng/mL
Oxymorphone	1.66	302.1	227.2, 198.2	25—2000 ng/mL
Hydromorphone	1.90	286.1	185.2, 157.2	25—2000 ng/mL
Codeine	2.40	300.2	215.1, 165.2	25—2000 ng/mL
6-Monoacetyl morphine	2.50	328.1	165.1, 211.1	2.5—200 ng/mL
Oxycodone	2.52	316.2	298.2, 241.2	25—2000 ng/mL
Hydrocodone	2.62	300.2	199.0, 171.0	25—2000 ng/mL
Norfentanyl	3.01	233.2	150.1, 84.1	2.5—200 ng/mL
Norbuprenorphine	3.75	414.3	101.3, 83.1	25—2000 ng/mL
Fentanyl	4.06	337.3	188.2, 105.1	2.5—200 ng/mL
EDDP (methadone metabolite)	4.12	278.2	234.2, 186.2	25—2000 ng/mL
Buprenorphine	4.45	468.3	396.2, 414.3	10—200 ng/mL
Methadone	4.58	310.0	105.2, 77.1	25—2000 ng/mL

Source of data: Yang HS, Wu AH, Lynch KL. Development and validation of a novel LC—MS/MS opioid confirmation assay: evaluation of β-glucuronidase enzymes and sample clean up methods. J Anal Toxicol. 2016;40:323—329.

nor-diazepam, desalkyl-flurazepam, temazepam and α-hydroxy-alprazolam by GC/MS using tert-butyldimethylsilyl derivatives. The authors used d5-oxazepam, d5-nor-diazepam, and d5-α-hydroxyalprazolam as internal standards [47]. LC—MS/MS can also be analyzed for the analysis of various benzodiazepines. Jeong et al. analyzed various benzodiazepines, zolpidem, and their metabolites using LC—MS/MS where urine specimens were mixed with deuterated internal standards, and after centrifugation, an aliquot was directly injected into LC—MS/MS [48].

Issues of Drug Metabolites in Drug Confirmation

For pain management drug testing using GC/MS or LC—MS/MS, it is important to also confirm the presence of metabolites because sometimes people try to beat drug testing by adding parent drug in vitro after urine collection. Obviously, it is easy to catch these cheaters because no drug metabolite should be present and also there is a possibility of the parent drug to be present in supraphysiologically high concentration. In Fig. 7.1, metabolite pathways of various opioids including illicit drug heroin are given. In Fig. 7.2, metabolic pathways of commonly prescribed benzodiazepines are presented. In pain management drug testing, both prescribed drug and the metabolite should be

confirmed indicating that the patient took the drug as prescribed.

Case Report

A patient was prescribed alprazolam, MS Contin (morphine sulfate-extended release), and oxycodone. Immunoassay screening was positive for opiates and oxycodone but negative for benzodiazepines. However, drug confirmation step using LC—MS/MS confirmed the presence of α-hydroxy-alprazolam (22 ng/mL), alprazolam (16 ng/mL), hydromorphone (74 ng/mL0, morphine (4990 ng/mL), oxycodone (6910 ng/mL, oxymorphone (1150 ng/mL), and nor-oxycodone (9950 ng/mL). The urine also showed the presence of hydrocodone (15 ng/mL and nor-hydrocodone (26 ng/mL). The detectable hydrocodone was consistent with pharmaceutic impurity in oxycodone formulation (tolerable limit up to 1%). Therefore, urine drug confirmation results are consistent with prescribed medication. However, if only immunoassay was used, the patient was wrongly accused of not taking alprazolam because alprazolam must be present at 219 ng/mL to trigger a positive immunoassay response. Therefore, alprazolam concentration was too low to trigger a positive response in the commercial immunoassay used for benzodiazepine screening [49]. Acceptable impurities present in commercial opioid drugs are listed in Table 7.6.

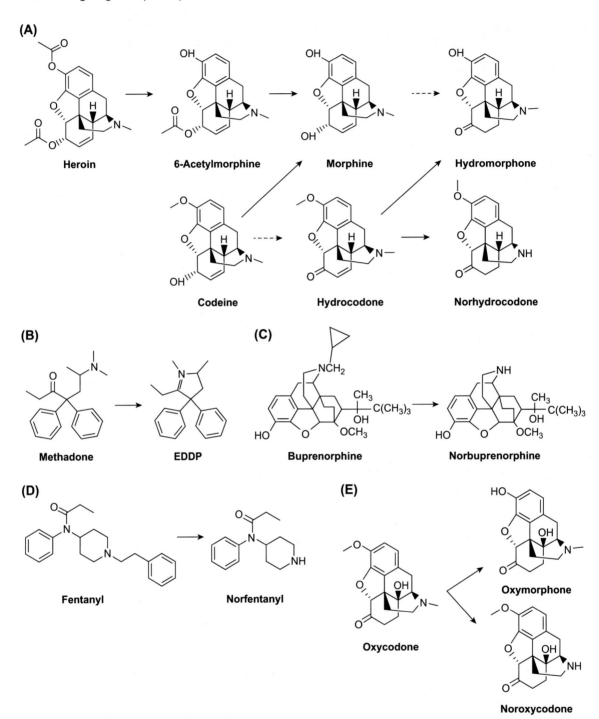

FIG. 7.1 Opiate drugs and common metabolites: (A) heroin and its metabolites, (B) methadone and its metabolites, (C) buprenorphine and its metabolites, (D) fentanyl and its metabolites, and (E) oxycodone and its metabolites. (Source: Sobolesky PM, Smith BE, Pesce A, Fitzgerald RL. Interpretation of pain management testing results using case examples. J Appl Lab Med. 2018;2(4):610–621; http://doi.org/10.1373/jalm.2017. 024786. Reproduced with permission from the American Association for Clinical Chemistry.)

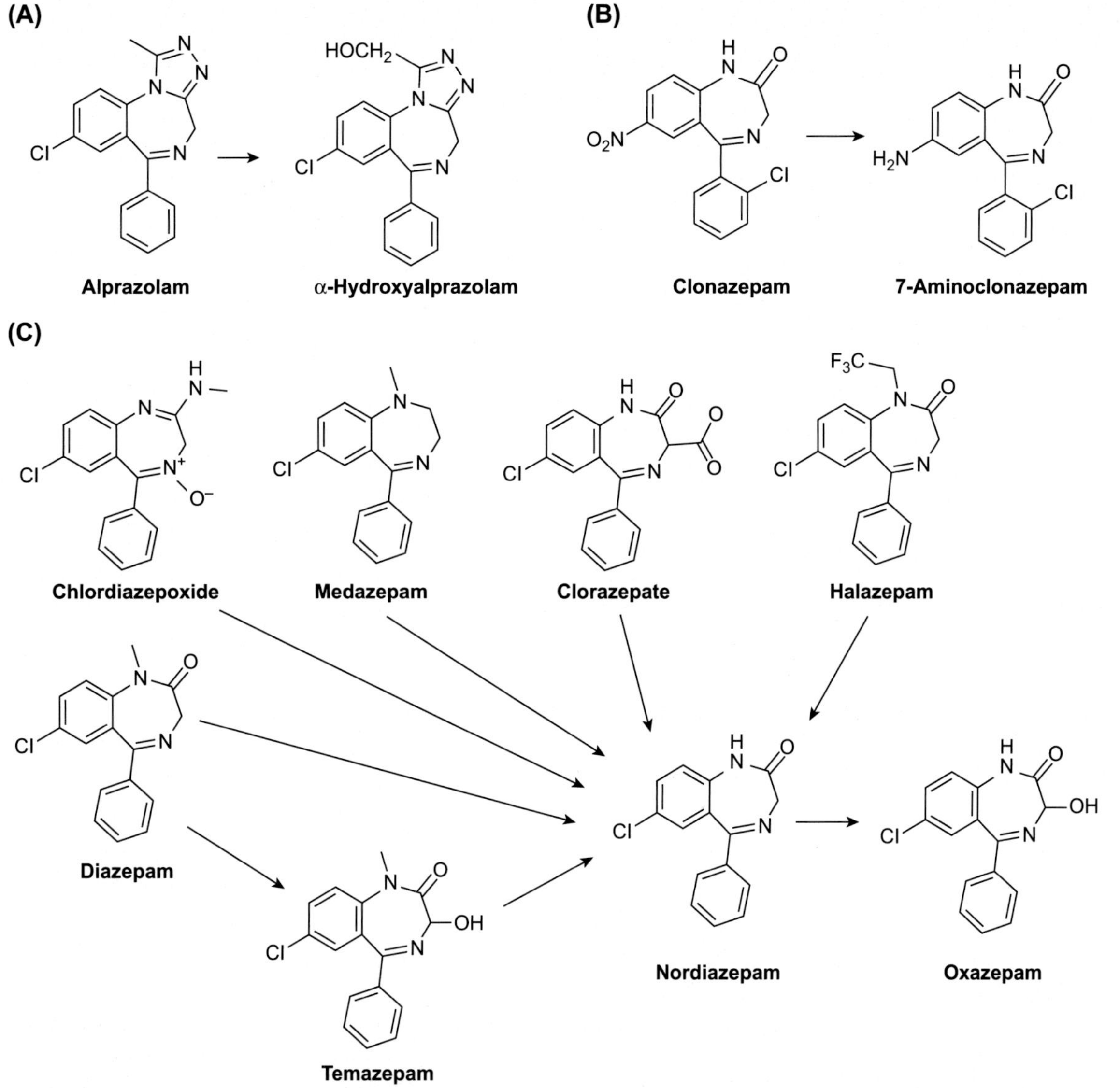

(A)

Alprazolam → α-Hydroxyalprazolam

(B)

Clonazepam → 7-Aminoclonazepam

(C)

Chlordiazepoxide   Medazepam   Clorazepate   Halazepam

Diazepam

Temazepam

Nordiazepam → Oxazepam

FIG. 7.2 Benzodiazepine drugs and their common metabolites. (Source: Sobolesky PM, Smith BE, Pesce A, Fitzgerald RL. Interpretation of pain management testing results using case examples. J Appl Lab Med. 2018; 2(4):610–621; http://doi.org/10.1373/jalm.2017.024786. Reproduced with permission from the American Association for Clinical Chemistry.)

Case Report

The urine of a patient admitted for chest and epigastric pain tested positive for cocaine using an immunoassay-based drug screening method (cutoff concentration 150 ng/mL). Despite the patient's denial of recent cocaine use, this positive cocaine screening result in conjunction with a remote history of drug misuse prompted the clinical team to question the appropriateness of opioids and other potentially addictive therapeutics during the treatment of cancer pain from previously undetected advanced pancreatic carcinoma. After pain management and clinical pathology consultation, it was decided that the positive cocaine screening result should be confirmed by GC/MS testing. This more sensitive and specific analytical technique revealed that both cocaine and its primary metabolite benzoylecgonine were undetectable (i.e., less than the assay detection limit of 50 ng/mL), thus indicating

TABLE 7.6
Acceptable Impurities Present in Commercial Opioid Drugs.

Opioid Drug	Impurity Present	Allowable Limit	Comments
Codeine	Morphine	0.15%	Morphine impurity <1%
Morphine	Codeine 6-Monoacetyl morphine	0.5% Not established	Codeine impurity <1% for
Hydrocodone	Codeine	0.15%	Newer preparations may not contain any impurity
Oxycodone	Hydrocodone	1%	Hydrocodone <0.2%
Oxymorphone	Hydromorphone Oxycodone	0.15%	Hydromorphone up to1% Oxycodone up to 0.5%
Hydromorphone	Hydrocodone Morphine	0.1% 0.15%	Newer preparation may not contain any impurity

Source: Medical Review Officer Alert, volume 21, number 3, Triangle Park, NC: Quadrangle Research, LLC, Research; April 2010. Inclusion of SAMHSA content does not constitute or imply endorsement or recommendation by the Substance Abuse and Mental Health Services Administration, the U.S. Department of Health and Human Services, or the U.S. Government.

that the positive urine immunoassay screening result was false positive. With this confirmation, the pain management service team was reassured in offering intrathecal pump (ITP) therapy for pain control and the patient eventually achieved excellent pain relief. However, ITP therapy most likely would not have been utilized without the GC/MS confirmation testing. The authors concluded that confirmatory drug testing should be performed on specimens with unexpected immunoassay-based drug screening results [50].

Case Report

The patient was prescribed Suboxone (buprenorphine and naloxone combination). The urine specimen screened positive by immunoassay for buprenorphine and benzoylecgonine (inactive metabolite of cocaine) but negative for naloxone. However, further analysis using LC–MS/MS confirmed the presence of naloxone in urine along with buprenorphine (15 ng/mL), nonbuprenorphine (20 ng/mL), and benzoylecgonine (16,700 ng/mL). The positive cocaine result as benzoylecgonine can be interpreted as illicit drug use [49].

ISSUES OF SPECIMEN ADULTERATION

In forensic urine drug testing, guidelines have been established for detecting tampering with a urine specimen. Acceptable ranges for temperature (32.5–37.8°C), pH (4.7–7.8), creatinine concentration (20 mg/dL or higher), and specific gravity (1.003–1.035) are intended to reveal specimens that have been substituted (temperature), adulterated, or diluted (creatinine and specific gravity). Substitution with another person's urine is also encountered in unobserved collections, and may

be a successful way to cheat on drug tests. Chemical warmers are sometimes used to heat a substituted specimen to a temperature within an acceptable range. In general, these specimen integrity checks are also appropriate for nonforensic urine drug testing, such as monitoring patients who are prescribed long-term opioid therapy, particularly when the screening is performed for the purpose of ensuring that the patient is not taking nonprescribed drugs.

People also try to beat drug tests by adding in vitro various household chemicals such as vinegar, lemon juice, soap, bleach, detergent, Visine eye drop papain/meat tenderizer, and a variety of other household chemicals and also Golden seal tea. In addition, various in vitro adulterants can also be obtained through Internet order or clandestine laboratories. These chemicals include potassium nitrite, pyridinium chlorochromate, glutaraldehyde, zinc sulfate, oxidants, and peroxides. Routine specimen integrity testing may not identify such adulterants but special dip sticks and other methods are available to detect such adulterants in urine before analysis. It is important to identify adulterated specimens before analysis because some of these adulterants can not only invalidate screening tests but may also interfere with GC/MS or LC–MS/MS confirmation test due to chemical degradation of certain drug metabolites [50,51].

Urine drug testing in pain management often serves the additional purpose of verifying adherence to prescribed opioid regimens; a negative screen for a prescribed drug may indicate that the drug has been used up early in the month and hence not administered in the days before the test, or that the drug has not been taken, but instead has been hoarded or diverted.

Ironically, a nonadherent patient who is tested for a prescribed drug has an incentive to ensure the urine drug testing is positive rather than negative (in contrast to drug abusers, whose incentive is to produce a negative specimen) and may not try to adulterate urine in vitro to invalidate drug testing. Two potential approaches to gain a positive result are the substitution of a urine specimen from someone who is taking the drug, or addition of the drug ("spiking") to the negative urine specimen. The latter approach has two important pitfalls:

- Concentrations of drugs in the urine typically fall within predictable limits although extremes are sometimes encountered. The addition of a few milligrams of the drug to a 50 mL urine specimen can produce a urinary drug concentration of more than 100,000 ng/mL, which is unusually high and inconsistent with prescription use of opioids.
- Drugs (opioids and benzodiazepines) are rarely eliminated exclusively as the parent drug. Most opioids undergo some type of oxidative metabolism, as well as conjugation. Therefore, the presence of solely the parent drug, or in some cases, concentrations of parent drug that greatly exceed metabolite(s), in a urine specimen would be suspicious, and metabolites are prohibitively difficult to obtain by patients who may attempt to cheat on a pain management urine drug test.

Interpreting Urine Drug Testing Results in Pain Management

Results of pain management urine drug testing results must be carefully interpreted taking into account complex metabolic patterns of opioids. Although immunoassay and point of care devices could be used for initial testing of urine specimens collected from pain management patients, if immunoassay results do not correlate with intake drug, confirmation step using GC/MS or LC−MS/MS must be conducted before making any further clinical decisions. Following scenarios may arise during the interpretation of test results:

- Urine drug screen is negative for opioids prescribed: This may indicate ordering the wrong test by clinicians, for example, ordering opiate screening tests to monitor compliance with oxycodone, oxymorphone, hydrocodone, and hydromorphone. These drugs cannot be detected by opiate immunoassays due to poor cross-reactivity with an antibody that targets morphine in opiate immunoassay. Therefore, the clinician should order opioids confirmation using GC/MS or LC−MS/MS to resolve this issue.

- Urine drug screen is negative and opioids confirmation test result is also negative: This is an indication of noncompliance or diversion of prescription opioids. If a patient does not have a valid explanation for this finding, the physician may consider taking action and not prescribing any more opioids.
- Urine drug screening positive for opioids and illicit drugs: Physicians must order drug confirmation to investigate if immunoassay results are false positive. If both prescription and nonprescription drugs are confirmed then the physician must discuss such results with the patient before taking further action.
- Immunoassay negative for opiates but positive for illicit drugs. The physician must order confirmation and if no opioid is confirmed but the presence of illicit drugs is confirmed then the patient is abusing illicit drugs and may not suffer from pain because he/she is not taking any pain medication.
- Opiate screen is positive but showing very high results: In this case, opioid confirmation should be ordered and the presence of very high amount of the parent drug but no drug metabolite indicates that the patient is trying to cheat the drug test by adding the drug in vitro.
- Urine creatinine is very low: In this case, most likely a patient is trying to cheat the drug test by adding prescribed drug in vitro and then adding water to lower the concentration in the physiological range. In addition, the absence of any opioid metabolite is a strong indication that the drug is added in vitro.

USE OF ORAL FLUID FOR PAIN MANAGEMENT DRUG TESTING

One major limitation of urine drug testing in pain management patients is the difficulty in interpreting test results due to the lack of correlation between blood levels of drugs and urine levels. Urine drug testing only confirms that the patient is taking prescribed opioid but is difficult to conclude based on urine levels whether the patient is taking more than prescribed dose or not. Because of this limitation, oral fluid testing is emerging as a useful, convenient, and reliable alternative matrix for drug testing. Oral fluid (also called saliva) is a mixture of fluids secreted from three major glands (parotid, sublingual, and submandibular) as well as food debris, bacteria, and very little protein. A drug concentration in oral fluid is a reflection of nonprotein bound (free drug) in circulation when equilibrium is reached. Usually, drug metabolites are absent or present in relatively low concentrations in the oral fluid due to their polar nature. Advantages in drug testing in oral fluid include ease of collection and direct observation so that a patient cannot tamper with the oral fluid specimen. In addition, medications often prescribed in pain management tend to accumulate well in oral fluid due to their basic nature because pH of oral fluid is slightly acidic. As a result, oxycodone, tramadol, hydrocodone, buprenorphine, methadone, and fentanyl are present in sufficient concentration for successful detection. However, benzodiazepines being acidic in nature are present in low concentrations in oral fluid. Another disadvantage of oral fluid drug testing is low specimen volume. Usually, the volume of oral fluid collected is 1 mL versus a typical urine volume of 30–50 mL. Moreover, the detection window for drugs in oral fluid is shorter than urine [52].

Coulter and Moore commented that oral fluid analysis for drugs is increasingly used in a variety of testing areas: pain management and medication monitoring, parole, and probation situations, driving under the influence of drugs, therapeutic drug monitoring, and testing for drugs in the workplace. The sample collection itself is straightforward, rapid, observable, and noninvasive, requiring no special facilities. The pH of saliva is slightly acidic relative to blood; therefore, drugs that are more basic tend to be present in higher concentration in the oral fluid than in blood: cocaine, amphetamines, oxycodone, tramadol, buprenorphine, methadone, and fentanyl. Conversely, acidic drugs and drugs that are strongly protein bound have lower concentrations in the oral fluid than in blood: examples include benzodiazepines, barbiturates, and carisoprodol. Because of the low volume of specimen available for analysis and the drug concentrations present (generally much lower than those in urine), efficient extraction methods and sensitive confirmation procedures such as liquid chromatography combined with mass spectrometry are needed for detection and confirmation of drugs in oral fluid specimens [53].

West et al. performed a retrospective review of paired oral fluid and urine test results from Millennium Health's laboratory database using 2746 patients with reported prescriptions for buprenorphine products. Specimens were tested using quantitative LC–MS/MS for 34 medications, metabolites, and illicit drugs. The authors detected a number of medications and illicit drugs at comparable or higher rates in oral fluid versus urine such as cocaine (15.7% vs. 7.9%), opiates (13.4% vs. 10.0%), oxycodone (8.6% vs. 3.7%), hydrocodone (3.0% vs. 1.2%), and others. However, lower detection rates were observed in oral fluid versus urine for benzodiazepines (6.6% vs. 8.7%), cannabinoids (15.5% vs. 19.5%), oxymorphone (1.8% vs. 3.1%),

and hydromorphone (0.8% vs. 4.5%). The authors concluded that clinicians may find oral fluid advantageous for the detection of specific drugs and medications in certain clinical situations. However, understanding the relative differences between urine and oral fluid can help clinicians to carefully select tests best suited for detection in their respective matrix [54].

CONCLUSIONS

Drug testing in pain management patients is recommended before initiation of therapy with opioids and then at least once a year to monitor compliance and or use of nonprescription opioids as well as illicit drugs. For pain management patients who may indicate a higher risk of misuse of opioids or abuse of illicit drugs, follow-up drug testing may be conducted twice or three times a year depending on the potential risk of abuse. Urine is the most common specimen for pain management drug testing. Although immunoassays are widely used for pain management drug testing, it is important to order the correct test, for example, if opiate immunoassay is ordered to monitor compliance with oxycodone, then the result will be false negative because oxycodone has a very poor cross-reactivity with antibodies utilized in immunoassays that are targeted to detect morphine. In addition, immunoassay results may be false negative even if correct immunoassay is used because immunoassays especially benzodiazepine immunoassays have inadequate sensitivity to detect certain benzodiazepines especially their conjugated metabolites. Moreover, immunoassay cut-off of certain opioids may not be adequate to detect such drugs in urine if prescribed in low dosage. Therefore, it is advisable not to dismiss any patient from pain clinic based on the immunoassay test result. The gold standard for pain management drug testing is GC/MS or LC–MS/MS.

REFERENCES

[1] Bush D. The US mandatory guidelines for Federal workplace drug testing programs: current status and future considerations. Forensic Sci Int 2008;174:111–9.

[2] Department of Health and Human Services. Proposed revision to mandatory guidelines for federal workplace drug testing programs. Fed Regist 2015;80:28101–51.

[3] Eichhorst JC, Etter ML, Rousseaux N, Lehotay DC. Drugs of abuse testing by tandem mass spectrometry: a rapid simple method to replace immunoassay. Clin Biochem 2009;42:1531–42.

[4] Dietzen DJ, Ecos K, Friedman D, Beason S. Positive predictive values of abused drug immunoassays on the Beckman SYNCHRON in a veteran population. J Anal Toxicol 2001;25:174–8.

[5] Casey ER, Scott MG, Tang S, Mullins ME. Frequency of false positive amphetamine screens due to bupropion using the Syva EMIT II immunoassay. J Med Toxicol 2011;7:105−8.

[6] Baron JM, Griggs DA, Nixon AL, Long WH, et al. The trazodone metabolite meta-chlorophenylpiperazine can cause false positive urine amphetamine immunoassay result. J Anal Toxicol 2011;35:364−8.

[7] Yee LM, Wu D. False positive amphetamine toxicology screen results in three pregnant women using labetalol. Obstet Gynecol 2011;117(2 Pt 2):503−6.

[8] Merigian KS, Beowning RG. Desipramine and amantadine causing false positive urine test for amphetamine. Ann Emerg Med 1993;22:1927−8.

[9] Vorce SP, Holler JM, Cawrse BM, Magluilo J. Dimethylamine: a drug causing positive immunoassay results for amphetamines. J Anal Toxicol 2011;35:183−7.

[10] Baden LR, Horowitz G, Jacoby H, Eliopoulos GM. Quinolones and false positive urine screening for opiates by immunoassay technology. J Am Med Assoc 2001;286:3115−9.

[11] De Paula M, Saiz LC, Gonzalez-Revalderia J, Pascual T, et al. Rifampicin causes false positive immunoassay results for opiates. Clin Chem Lab Med 1998;36:241−3.

[12] Widschwendter CG, Zernig G, Hofer A. Quetiapine cross-reactivity with urine methadone immunoassays. Am J Psychiatr 2007;164:172.

[13] Rogers SC, Pruitt CW, Crouch DJ, Caravati EM. Rapid urine drug screens: diphenhydramine and methadone cross-reactivity. Pediatr Emerg Care 2010;26:665−6.

[14] Hausmann E, Kohl B, von Boehmer H, Wellhoner HH. False positive EMIT indication for opiates and methadone in doxylamine intoxication. J Clin Chem Clin Biochem 1983;21:599−600.

[15] Lichtenwalner MR, Mencken T, Tully R, Petosa M. False positive immunochemical screen for methadone attributable to metabolites of verapamil. Clin Chem 1998;44:1039−2041.

[16] Boucher A, Vilette P, Crassard N, Bernard N, et al. Urinary toxicological screening: analytical interference between niflumic acid and cannabis. Arch Pediatr 2009;16:1457−60 [Article in French].

[17] Oosthuizen NM, Laurens JB. Efavirenz interference in urine screening immunoassays for tetrahydrocannabinol. Ann Clin Biochem 2012;49:194−6.

[18] Rollins DE, Jennison TA, Jones G. Investigation of interference by nonsteroidal anti-inflammatory drugs in urine tests for abused drugs. Clin Chem 1998;36:602−6.

[19] Marchei E, Pellegrini M, Pichini S, Martin I, et al. Are false positive phencyclidine immunoassay instant-view multi test results caused by overdose concentrations of ibuprofen, metamizol and dextromethorphan? Ther Drug Monit 2007;29:671−3.

[20] Long C, Crifasi J, Maginn D. Interference of thioridazine (Mellaril) in identification of phencyclidine. Clin Chem 1996;42:1885−6.

[21] Bond GR, Steele PE, Uges DR. Massive venlafaxine overdose resulted in a false positive Abbott AxSYM urine immunoassay for phencyclidine. J Toxicol Clin Toxicol 2003;41:999−1002.

[22] Fraser AD, Howell P. Oxaprozin cross-reactivity in three commercial immunoassays for benzodiazepines in urine. J Anal Toxicol 1998;22:50−4.

[23] Nasky KM, Cowan GL, Knittel DR. False positive urine screening for benzodiazepines: an association with streamline?: a two year retrospective chart analysis. Psychiatry 2009;6:36−9.

[24] Maurer HH. Perspectives of liquid chromatography coupled to low- and high-resolution mass spectrometry for screening, identification, and quantification of drugs in clinical and forensic toxicology. Ther Drug Monit 2010;32:324−7.

[25] Tsai IL, Weng TI, Tseng YJ, Tan HK, et al. Screening and confirmation of 62 drugs of abuse and metabolites in urine by ultra-high-performance liquid chromatography-quadrupole time-of-flight mass spectrometry. J Anal Toxicol 2013;37:642−51.

[26] O'Byrne PM, Kavanagh PV, McNamara SM, Stokes SM. Screening of stimulants including designer drugs in urine using a liquid chromatography tandem mass spectrometry system. J Anal Toxicol 2013;37:64−73.

[27] Arthur JA. Urine drug testing in cancer pain management. Oncologist 2019;24:1−6.

[28] Seth P, Scholl L, Rudd RA, Bacon S. Overdose deaths involving opioids, cocaine, and psychostimulants − United States, 2015−2016. Morb Mortal Wkly Rep 2018;67(12):349−58.

[29] Urine drug testing CDC Guidelines. Available from: https://www.cdc.gov/drugoverdose/pdf/prescribing/CDC-DUIP-UrineDrugTesting_FactSheet [Accessed 3 December 2019].

[30] Argoff CE, Alford DP, Fudin J, Adler JA, et al. Rational urine drug monitoring in patients receiving opioids for chronic pain: consensus recommendations. Pain Med 2018;19:97−117.

[31] Michna E, Ross EL, Hynes WL, Nedeljkovic SS, et al. Predicting aberrant drug behavior in patients treated for chronic pain: importance of abuse history. J Pain Symptom Manag 2004;28:250−8.

[32] Berndt S, Maier C, Schütz HW. Polymedication and medication compliance in patients with chronic non-malignant pain. Pain 1993;52:331−9.

[33] Fishbain DA, Cutler RB, Rosomoff HL, Rosomoff RS. Validity of self-reported drug use in chronic pain patients. Clin J Pain 1999;15:184−91.

[34] Michna E, Jamison RN, Pham LD, Ross EL, et al. Urine toxicology screening among chronic pain patients on opioid therapy: frequency and predictability of abnormal findings. Clin J Pain 2007;23:173−9.

[35] Pesce A, West C, Gonzales E, Rosenthal M, et al. Illicit drug use correlates with negative urine drug test results for prescribed hydrocodone, oxycodone, and morphine. Pain Physician 2012;15:E687−92.

[36] Dixon RB, Davis B, Dasgupta A. Comparison of response of DRI oxycodone semiquantitative immunoassay with true oxycodone values determined by liquid chromatography combined with tandem mass spectrometry: sensitivity of the DRI assay at 100 ng/ml cut-off and validity of semiquantitative value. J Clin Lab Anal 2016;30:190−5.

[37] Dixon RB, Dasgupta A. Suitability of the DRI hydroco-
done/hydromorphone immunoassay in the clinical environment at a lower cutoff: validation with LC-MS/MS
analysis. Ther Drug Monit 2016;38(6):787—90.

[38] Mikel C, Almazan P, West R, Crews B, et al. Ther Drug
Monit 2009;31:746—8.

[39] Goggin MM, Shahriar BJ, Stead A, Janis GC. Reduced urinary opioid levels from pain management patients associated with marijuana use. Pain Manag 2019;9:441—7.

[40] Von Seggern RL, Fitzgerald CP, Adelman LC, Adelman JU.
Laboratory monitoring of OxyContin (oxycodone): clinical pitfalls. Headache 2004;44:44—7.

[41] West R, Pesce A, West C, Crews B, et al. Comparison of
clonazepam compliance by measurement of urinary
concentration by immunoassay and LC-MS/MS in pain
management population. Pain Physician 2010;13:71—8.

[42] Darragh A, Snyder ML, Ptolemy AS, Melanson S. KIMS,
CEDIA, and HS-CEDIA immunoassays are inadequately
sensitive for detection of benzodiazepines in urine from
patients treated for chronic pain. Pain Physician 2014;
17:359—66.

[43] Broussard L, Presley LC, Pittaman T, Clouette R, et al.
Simultaneous identification and quantitation of codeine,
morphine, hydrocodone and hydromorphone in urine
as trimethylsilyl and oxime derivatives by gas
chromatography-mass spectrometry. Clin Chem 1997;
43:1029—32.

[44] Cone EJ, Zichterman A, Heltsley R, Black DL, et al. Urine
testing for nor-codeine, nor-hydrocodone and noroxycodone facilitates interpretation and reduces false negatives.
Forensic Sci Int 2010;198:58—61.

[45] Yang HS, Wu AH, Lynch KL. Development and validation
of a novel LC-MS/MS opioid confirmation assay: evaluation of beta-glucuronidase enzymes and sample clean up
methods. J Anal Toxicol 2016;40:323—9.

[46] Valentine JL, Middleton R, Sparks C. Identification of urinary benzodiazepines and their metabolites: comparison
of automated HPLC, GC-MS after immunoassay screening
of clinical specimens. J Anal Toxicol 1996;20:416—24.

[47] West RE, Ritz DP. GC/MS analysis of for five common
benzodiazepine metabolites in urine as tert-
butyldimethylsilyl derivatives. J Anal Toxicol 1993;17:
114—6.

[48] Jeong YD, Kim MK, Suh SI, In MK, et al. Rapid determination of benzodiazepines, zolpidem and their metabolites
in urine using direct injection liquid chromatography-
tandem mass spectrometry. Forensic Sci Int 2015;257:
84—92.

[49] Sobolesky PM, Smith BE, Pesce A, Fitzgerald RL. Interpretation of pain management testing results using case
examples. J Appl Lab Med 2018;2:610—21.

[50] Kim JA, Ptolemy AS, Melanson SE, Janfaza DR, Ross EL.
The clinical impact of a false-positive urine cocaine
screening result on a patient's pain management. Pain
Med 2015;16:1073—6.

[51] Fu S. How do people try to beat drug test? Effects of synthetic urine, substituted urine, and in vitro urinary adulterants on drugs of abuse testing [Chapter 26]. In:
Dasgupta A, editor. Critical issues in alcohol and drugs
of abuse testing. Elsevier; 2019. p. 359—89.

[52] Kwong TC, Magnani B, Moore C. Urine and oral fluid
drug testing in support of pain management. Crit Rev
Clin Lab Sci 2017;54:433—45.

[53] Coulter CA, Moore CM. Analysis of drugs in oral fluid using LC-MS/MS. Methods Mol Biol 2019;1872:237—59.

[54] West R, Mikel C, Hofilena D, Guevara M. Positivity rates
of drugs in patients treated for opioid dependence with
buprenorphine: a comparison of oral fluid and urine using paired collections and LC-MS/MS. Drug Alcohol
Depend 2018;193:183—91.

Index

Note: Page numbers followed by "f" indicate figures and "t" indicates tables.

Printed in the United States
By Bookmasters